The book I wish I'd had when I stepped up to principal engineer. If you are wondering what Staff+ means, and how to be successful in the role in your organization, Tanya has laid out the path, with lots of practical, insightful advice for you. This book will provide the tools to enable you to thrive in a tech track role, acting with influence and impact.

—*Sarah Wells, independent consultant and author,*
former principal engineer at the Financial Times

This book feels like the missing manual for my whole career. It's amazingly reassuring to see the ambiguity of the role laid out in print, along with great specific guidance on time management, consensus building, etc. I'm going to cite this a lot.

—*Titus Winters, principal engineer, Google, and*
coauthor of Software Engineering at Google

Tanya is the perfect author for this exceptional guide to navigating the murky role of staff-plus engineering. Her deep, direct experience comes through in every section and taught me a great deal.

—*Will Larson, CTO, Calm, and author of* Staff Engineer

The job of senior leadership as an individual contributor has long been ambiguous and difficult to define, and this book is a much-needed guide on being successful in a relatively new role to our industry. Tanya does an excellent job bringing large-company perspective and scaling company challenges for a rounded view on how to be a successful staff engineer.

—*Silvia Botros, principal engineer and*
coauthor of High Performance MySQL, *4th edition*

When you reach near the top of the individual-contributor scale, you're given a metaphorical compass and a destination. How you get there is your problem. How you lead there is everybody's problem. Tanya offers a solid framework, a mapping approach, to help you lead from "here" to "there." This book offers a solid anchor for those new to the upper levels of individual contributors, and new perspectives for those with more experience. Staff engineer, know thyself.

—Izar Tarandach, principal security architect and coauthor of Threat Modeling

Tanya Reilly captures with eerie accuracy the sinking feeling I experienced when I first became the "someone" in "someone should do something." This book is a detailed exploration of what that actually means for folks at the staff engineer level.

—Niall Richard Murphy, founder, CEO, and coauthor of Reliable Machine Learning and Site Reliability Engineering

In The Staff Engineer's Path, Tanya Reilly has brought desperately needed clarity to the ambiguous and often misunderstood question of how to be a senior technical leader without direct reports. Every page is chock full of valuable insights and actionable advice for navigating your role, your org, and carving out your career path—all delivered in Tanya's trademark witty, insightful, and down-to-earth style. This book is a masterpiece.

—Katie Sylor-Miller, senior staff frontend architect, Etsy

If you're a senior engineer wondering what the next level is—a staff-level engineer or a manager of staff engineers—this book is for you. It covers so many of the things no one tells you about this role— things that take long years, even with great mentors, to discover on your own. It offers observations, mental models, and firsthand experiences about the staff engineer role in a more distilled way than any other book has covered before.

—Gergely Orosz, author of The Pragmatic Engineer

The Staff
Engineer's Path

A Guide for Individual Contributors
Navigating Growth and Change

Tanya Reilly

Beijing · Boston · Farnham · Sebastopol · Tokyo

The Staff Engineer's Path

by Tanya Reilly

Copyright © 2022 Tanya Reilly. All rights reserved.

Published by O'Reilly Media, Inc., 1005 Gravenstein Highway North, Sebastopol, CA 95472.

O'Reilly books may be purchased for educational, business, or sales promotional use. Online editions are also available for most titles (*http://oreilly.com*). For more information, contact our corporate/institutional sales department: 800-998-9938 or *corporate@oreilly.com*.

Acquisitions Editor: Melissa Duffield

Development Editor: Sarah Grey

Production Editor: Elizabeth Faerm

Copyeditor: Josh Olejarz

Proofreader: Liz Wheeler

Indexer: Sue Klefstad

Interior Designer: Monica Kamsvaag

Cover Designer: Susan Thompson

Cover Art Creator: Susan Thompson

Illustrator: Kate Dullea

September 2022:   First Edition

Revision History for the First Edition

2022-09-20   :   First Release

2023-01-06   :   Second Release

See *http://oreilly.com/catalog/errata.csp?isbn=9781098118730* for release details.

978-1-098-11873-0

[LSI]

Contents

Foreword

When I wrote *The Manager's Path* in 2016, I had many goals. I wanted to share lessons I had learned growing up as a manager. I wanted to show those who were interested in becoming managers what the job would be like. And I wanted to force a reckoning across the industry that we needed to expect more from our managers, and that the managers we were currently promoting often did not have the right balanced focus of people, process, product, and technical skills to do the job well. In short, I wanted to correct what I saw as a cultural failing in tech: to both take management seriously as a critical role and to discourage it from being the default path for ambitious engineers who want to grow their careers.

I would say that I partially succeeded. Every time someone tells me they read my book and decided not to become a manager, I do a little victory dance. From that perspective, at least some people read my book and realize that this path isn't for them. Unfortunately, the alternative path of career growth for the individual contributor, the staff+ engineering path, has lacked a similar guidebook. This lack has led to many choosing to follow the management path despite knowing they would rather not have the responsibility for larger and larger groups of people, because they cannot see another way forward. This is a great frustration for engineers and managers alike: most managers want to have more strong staff+ engineers in their organizations but don't know how to cultivate them, and many engineers want to stay on that path but see no realistic options beyond going into management.

One of the core challenges of the staff+ engineering path is the unspoken expectation that part of being qualified to be on that path is figuring out how to climb it without much in the way of directions. If you were destined to be a staff+ engineer, conventional wisdom argues, you would figure out how to get there yourself. Needless to say, this is a frustrating and bias-ridden approach to career development. As more and more companies realize the need for staff+ engineers, we as an industry cannot afford to maintain a mysticism about the staff+ engineering career path that ignores the underlying skills that lead to successful technical leaders.

With this in mind, you can imagine how thrilled I felt when Tanya Reilly came up with a proposal for a book about career growth as a staff engineer, filling in the missing half of the career ladder that my book left unexplored. I know Tanya from her writing and speaking about technical leadership, and it is obvious that she wants to correct the cultural failings of the tech industry's approach to staff+ engineering in the same way that I wanted to correct the cultural failing in the tech industry's approach to management. Namely, Tanya wants to address the overfocus on coding and technical contributions and the lack of clarity around the skills that allow strong engineers to become successful multipliers without needing to manage people.

In this book, Tanya has stepped up to the task of articulating these underlying skills that are so crucial for successful staff+ engineers. She provides a framework that shows how, using the pillars of big-picture thinking, execution, and leveling up others, you can build impact that goes beyond your individual hands-on contributions.

Reflecting the multifaceted nature of the staff+ engineering path, Tanya does not try to dictate the precise mix of skills needed at each level above senior engineer. Instead, she wisely focuses on how to build out these pillars from wherever you are today. From developing technical strategy to leading big projects successfully and going from mentor to organizational catalyst, this book takes you through these critical pillars and shows how to increase your impact on the success of your company beyond writing code.

There's only one person in the driver's seat for your career, and that person is you. Figuring out your career path is one of the great opportunities and challenges of your life, and the earlier you accept that it's up to you (plus a heaping dose of luck), the better set you are to navigate the working world. This guidebook shows you the skills you'll need on the staff+ path, and it's an essential addition to every engineer's library.

—Camille Fournier
Author, The Manager's Path
Editor, 97 Things Every Engineering Manager Should Know
Managing Director, JP Morgan Chase
Board Member, ACM Queue
New York, NY, September 2022

Introduction

Where do you see yourself in five years? The classic interview question is the adult equivalent of "What do you want to be when you grow up?": it has some socially acceptable answers and a long enough time horizon that you don't need to commit.[1] But if you're a senior software engineer looking to keep growing in your career, the question becomes very real.[2] Where *do* you see yourself going?

Two Paths

You may find yourself at a fork in the road (Figure P-1), two distinct paths stretching ahead. On one, you take on direct reports and become a manager. On the other, you become a technical leader without reports, a role often called *staff engineer*. If you really could see five years ahead on both of these paths, you'd find that they have a lot in common: they lead to many of the same places, and the further you travel, the more you'll need many of the same skills. But, at the start, they look quite different.

---

1 Although you're not supposed to reply "a zookeeper who is also an astronaut" to the interview question. Adult life is very limiting.

2 For the sake of brevity, I'm going to say "software engineer" throughout this book; however, if you're a systems engineer, data scientist, or any other practitioner of tech, I think you'll find it relevant, too. All are welcome here!

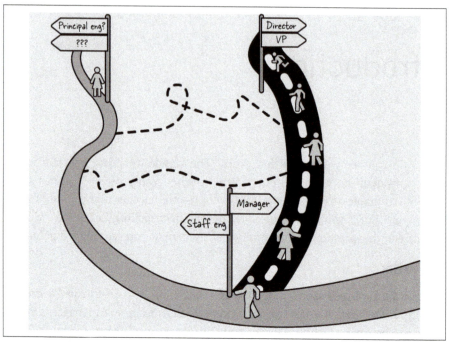

Figure P-1. A fork in the road.

The manager's path is clear and well traveled. Becoming a manager is a common, and perhaps default, career step for anyone who can communicate clearly, stay calm during a crisis, and help their colleagues do better work. Most likely, you know people who have chosen this path. You've probably had managers before, and perhaps you have opinions about what they did right or wrong. Management is a well-studied discipline, too. The words *promotion* and *leadership* are often assumed to mean "becoming someone's boss," and airport bookshops are full of advice on how to do the job well. So, if you set off down the management path, it won't be an *easy* road, but you'll at least have some idea of what your journey will be like.

The staff engineer's path is a little less defined. While many companies now allow engineers to keep growing in seniority without taking on reports, this "technical track" is still muddy and poorly signposted. Engineers considering this path may have never worked with a staff engineer before, or might have seen such a narrow set of personalities in the role that it seems like unattainable wizardry. (It's not. It's all learnable.) The expectations of the job vary across companies, and, even within a company, the criteria for hiring or promoting staff engineers can be vague and not always actionable.

Often the job doesn't become clearer once you're in it. Over the last few years, I've spoken with staff engineers across many companies who weren't quite sure what was expected of them, as well as engineering managers who didn't know how to work with their staff engineer reports and peers.[3] All of this ambiguity can be a source of stress. If your job's not defined, how can you know whether you're doing it well? Or doing it at all?

Even when expectations are clear, the road to achieving them might not be. As a new staff engineer, you might have heard that you're expected to be a technical leader, make good business decisions, and influence without authority—but *how?* Where do you start?

The Pillars of Staff Engineering

I understand that feeling. Through 20 years in the industry, I've stayed on the staff engineer's path, and I'm now a senior principal engineer, parallel to a senior director on my company's career ladder. While I've considered the manager's path many times, I've always concluded that the "technical track" work is what gives me energy and makes me want to come to work in the morning. I want to have time to dig into new technologies, deeply understand architectures, and learn new technical domains. You get better at whatever you spend time on, and I've consistently wanted to keep getting better at technical things.[4]

Earlier in my career, though, I struggled to make sense of this path. As a midlevel engineer, I didn't understand why we had levels above "senior"—what did those people *do* all day? I certainly couldn't see a path to those roles from where I was. Later, as a new staff engineer, I discovered unspoken expectations and missing skills I didn't know how to *describe*, much less act on. Over the years, I've learned from many projects and experiences—both successes and failures— as well as from phenomenal colleagues and peers in other companies. The job makes sense now, but I wish I'd known then what I know now.

If you've taken the staff engineer path, or are considering it, welcome! This book is for you. If you work with a staff engineer, or manage one, and want to

---

3 This is changing. Will Larson (*https://staffeng.com*), LeadDev (*https://leaddev.com/staffplus-new-york*), and others have been doing phenomenal work in paving the road. I'll link to resources throughout this book.

4 I reserve the right to change my mind later on.

know more about this emerging role, there'll be a lot here for you too. In the next nine chapters I'm going to share what I've learned about how to be a great staff engineer. I'll warn you right now that I'm not going to be prescriptive about every topic or answer every question: a great deal of the ambiguity is inherent to the role, and the answer is very often "it depends on the context." But I'll show you how to navigate that ambiguity, understand what's important, and stay aligned with the other leaders you work with.

I'll unpack the staff engineer role by looking at what I think of as its three pillars: *big-picture thinking, execution* of projects, and *leveling up* the engineers you work with.

Big-picture thinking

Big-picture thinking means being able to step back and take a broader view. It means seeing beyond the immediate details and understanding the context that you're working in. It also means thinking beyond the current time, whether that means initiating yearlong projects, building software that will be easy to decommission, or predicting what your company will need in three years.[5]

Execution

At the staff level, the projects you take on will become messier and more ambiguous. They'll involve more people and need more political capital, influence, or culture change to succeed.

Leveling up

Every increase in seniority comes with more responsibility for raising the standards and skills of the engineers within your orbit, whether that's your local team, colleagues in your organization, or engineers across your whole company or industry. This responsibility will include intentional influence through teaching and mentoring, as well as the accidental influence that comes from being a role model.

We can think of these three pillars as supporting your impact like in Figure P-2.

---

5 Throughout this book I'm going to use the term "company" when talking about employers, but of course you could be at a nonprofit, government agency, academic institution, or other type of organization. Swap in whatever makes sense for you.

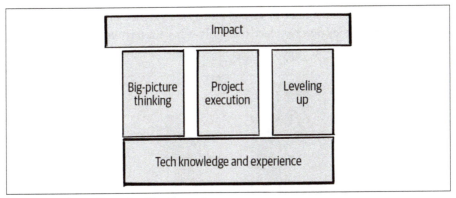

Figure P-2. Three pillars of staff engineer roles.

You'll notice that these pillars sit on a solid foundation of technical knowledge and experience. This foundation is critical. Your big-picture perspective includes understanding what's possible and having good judgment. When executing on projects, your solutions will need to actually solve the problems they set out to solve. When acting as a role model, your review comments should make code and designs better, and your opinions need to be well thought out—you need to be right! Technical skills are the foundation of every staff engineer role, and you'll keep exercising them.

But technical knowledge is not enough. Success and growth at this level means doing more than you can do with technical skills alone. To become adept at big-picture thinking, execute on bigger projects, and level up everyone around you, you're going to need "humaning" skills, like:

- Communication and leadership
- Navigating complexity
- Putting your work in perspective
- Mentorship, sponsorship, and delegation
- Framing a problem so that other people care about it
- Acting like a leader whether you feel like one or not[6]

---

6 And a lot more. Check out Camille Fournier's article "An Incomplete List of Skills Senior Engineers Need, Beyond Coding" (*https://oreil.ly/gGe2T*).

Think of these skills like the flying buttresses you see on gothic cathedrals (as in Figure P-3): they don't replace the walls—or your technical judgment—but they allow the architect to build taller, grander, more awe-inspiring buildings.

Figure P-3. Leadership skills are like the flying buttresses that let us keep massive buildings stable.

Each of the three pillars has a set of required skills, and your aptitude for each of them will vary. Some of us may be in our element when leading and finishing big projects, but find it intimidating to choose between two strategic directions. Others may have strong instincts for understanding where the company and industry are going, but lose control of the room quickly when managing an incident. Still others may boost the skills of everyone they work with, but struggle to build consensus around a technical decision. The good news is that all of these skills are learnable, and you can become adept at all three pillars.

This book is divided into three parts.

Part I: The Big Picture

In Part I, we'll look at how to take a broad, strategic view when thinking about your work. Chapter 1 will begin by asking big questions about your role. What's expected of you? What are staff engineers *for*? In Chapter 2, we'll zoom out further and get some perspective. We'll look at your work in context, navigate your organization, and uncover what your goals are. Finally, in Chapter 3, we'll look at adding to the big picture by creating a technical vision or strategy.

Part II: Execution

Part II gets tactical and moves on to the practicalities of leading projects and solving problems. In Chapter 4, we'll look at choosing what to work on: I'll share techniques for how to decide what to spend time on, how to manage your energy, and how to "spend" your credibility and social capital in a way that doesn't diminish it. In Chapter 5 I'll discuss how to lead projects that stretch across teams and organizations: setting them up for success, making the right decisions, and keeping information flowing. Chapter 6 will look at navigating the obstacles you'll meet along the way, celebrating a project that finishes successfully, and retrospecting (but still celebrating!) if it's canceled and cleanly shut down.

Part III: Leveling Up

Part III is about leveling up your organization. Chapter 7 will look at raising everyone's game by modeling what a great engineer acts like, how to learn out loud, and how to build a psychologically safe culture. We'll look at how to be the "adult in the room" during an incident or a technical disagreement. Chapter 8 is about more intentional forms of raising your colleagues' skills, like teaching and coaching, design review, code review, and making cultural change. Finally, Chapter 9 will explore how to level up *yourself*: how to keep growing and how to think about your career. Where do you go after your current role? I'll discuss some options.

One warning before we go further: this is a book about staying on the technical *track*. It is not a technical *book*. As I've said, you need a solid technical foundation to become a staff engineer. This book won't help you get that. Technical skills are domain-specific, and if you're here, I'm assuming that you already have —or are setting out to learn—whatever specialized skills you need in order to be one of the most senior engineers in your domain. Whether "technical" for you means coding, architecture, UX design, data modeling, production operations, vulnerability analysis, or anything else, almost every domain has a plethora of books, websites, and courses that will support you.

If you're someone who thinks that technical skills are the only ones that matter, you're unlikely to find what you're looking for in here. But, ironically, you might also be the person who'll get the most from this book. No matter how deep or arcane your technical knowledge, you'll find that work gets less annoying when you can persuade other people to adopt your ideas, level up the engineers around you, and breeze through the organizational gridlock that slows everyone down. Those skills aren't easy to learn, but I promise they're all learnable, and I'll do my best in this book to show the way.

Do you want to be a staff engineer? It's fine not to aspire to more senior engineering roles. It's also fine to move to the manager track (or go back and forth!), or to stay at the senior level, doing work you enjoy. But if you like the idea of helping achieve your organization's goals and continuing to build technical muscle while making the engineers around you better at their craft, then read on.

O'Reilly Online Learning

O'REILLY® For more than 40 years, *O'Reilly Media* has provided technology and business training, knowledge, and insight to help companies succeed.

Our unique network of experts and innovators share their knowledge and expertise through books, articles, and our online learning platform. O'Reilly's online learning platform gives you on-demand access to live training courses, in-depth learning paths, interactive coding environments, and a vast collection of text and video from O'Reilly and 200+ other publishers. For more information, visit *https://oreilly.com*.

How to Contact Us

Please address comments and questions concerning this book to the publisher:

O'Reilly Media, Inc.

1005 Gravenstein Highway North

Sebastopol, CA 95472

800-998-9938 (in the United States or Canada)

707-829-0515 (international or local)

707-829-0104 (fax)

We have a web page for this book, where we list errata, examples, and any additional information. You can access this page at *https://oreil.ly/staff-eng-path*.

Email *bookquestions@oreilly.com* to comment or ask technical questions about this book.

For news and information about our books and courses, visit *https://oreilly.com*.

Find us on LinkedIn: *https://linkedin.com/company/oreilly-media*.

Follow us on Twitter: *https://twitter.com/oreillymedia*.
Watch us on YouTube: *https://youtube.com/oreillymedia*.

Acknowledgments

Thank you to the many, many people who helped make this book a reality.

Thank you to Sarah Grey, best of all possible development editors, and all of the other phenomenal folks at O'Reilly, including acquisitions editor, Melissa Duffield; production editor, Liz Faerm; copy editor, Josh Olejarz; Susan Thompson, who created the incredible cover; and illustrator, Kate Dullea, who translated my pencil scrawls into gorgeous art. This was my first time writing a book, and you all took the terror out of it.

My thanks to Will Larson for his encouragement and support, and for helping the staff engineering community find each other for the first time. Also thanks to Lara Hogan for enthusiasm and introductions when I turned up in her DMs all "but could *I* write a book??" Thank you both for showing what sponsorship looks like.

I was lucky enough to have two of the wisest and most insightful engineers I know along on this journey. Cian Synnott and Katrina Sostek, this book is infinitely better for your review and feedback over the last year. In particular, I am indebted for your thoughtful suggestions around the parts that didn't work. Constructive criticism is always harder, and I'm grateful for your time and energy.

Many people generously shared their time to discuss ideas, offer feedback, or teach me something. I want to particularly thank Franklin Angulo, Jackie Benowitz, Kristina Bennett, Silvia Botros, Mohit Cheppudira, John Colton, Trish Craine, Juniper Cross, Stepan Davidovic, Tiarnán de Burca, Ross Donaldson, Tess Donnelly, Tom Drapeau, Dale Embry, Liz Fong-Jones, Camille Fournier, Stacey Gammon, Carla Geisser, Polina Giralt, Tali Gutman, Liz Hetherston, Mojtaba Hosseini, Cate Huston, Jody Knower, Robert Konigsberg, Randal Koutnik, Lerh Low, Kevin Lynch, Jennifer Mace, Glen Mailer, Keavy McMinn, Daniel Micol, Zach Millman, Sarah Milstein, Isaac Perez Moncho, Dan Na, Katrina Owen, Eva Parish, Yvette Pasqua, Steve Primerano, Sean Rees, John Reese, Max Schubert, Christina Schulman, Patrick Shields, Joan Smith, Beata Strack, Carl Sutherland, Katie Sylor-Miller, Izar Tarandach, Fabianna Tassini, Elizabeth Votaw, Amanda Walker, and Sarah Wells. Also, thanks to the many (so many!) other people I spoke with via DMs, email, hallway conversation, or in spirited Slack threads. You made this book better, and I appreciate you.

Thank you to the people who drink afternoon tea: you demonstrate the power of community every day. Thanks to everyone on the #staff-principal-engineering channel on Rands Leadership Slack for being relentlessly supportive and for sharing your experiences with humility. Huge appreciation to my colleagues at Squarespace and to the Google SRE diaspora. I've learned a ton from you all. And I want to thank Ruth Yarnit, Rob Smith, Mariana Valette, and the whole Lead Dev crew for the incredible technical leadership content they've shared with the world. Thanks for what you do.

Thank you to the Hillfolks, including that very good dog. I'm lucky and privileged to have you as friends. Thanks for letting me write in your caravan (and quarantine there when I had COVID!) I look forward to decades more of friendship and watching your baby oak trees grow.

Finally, to my whole family—my parents, Danny and Kathleen, and the whole extended clan—thank you for being patient over the last year while I fell off the planet.

And of course, Joel and Ms 9! I look forward to seeing you on Saturdays again. To Joel (who came up with the idea that humaning skills are "flying buttresses"), thank you for great conversations about engineering organizations and making good software. And thank you for all the sandwiches. And to Ms 9 (who was Ms 6 when I wrote the first draft of this book!): thank you for your excellent ideas, drawings, and hugs. I appreciate you lummoxes.

PART I

The Big Picture

What Would You Say You Do Here?

The idea of a staff engineer track, or "technical track", is new to a lot of companies. Organizations differ on what attributes they expect of their most senior engineers and what kind of work those engineers should do. Although most agree that, as Silvia Botros has written (*https://oreil.ly/xwgRn*), the top of the technical track is not just "more-senior seniors," we don't have a shared understanding of what it *is*. So we'll start this chapter by getting existential: why would an organization *want* very senior engineers to stick around? Then, armed with that understanding, we'll unpack the role: its technical requirements, its leadership requirements, and what it means to work autonomously.

Staff engineering roles come in a lot of shapes. There are many valid ways to do the job. But some shapes will be a better fit for some situations, and not all organizations will need all kinds of staff engineers. So I'll talk about how to characterize and describe a staff engineering role: its scope, depth, reporting structure, primary focus, and other attributes. You can use these descriptions to be precise about how you want to work, what kind of role you're looking to grow into, or who you need to hire. Finally, since different companies have different ideas of what a staff engineer should do, we'll work on aligning your understanding with that of other key people in your organization.

Let's start with what this job even is.

What Even Is a Staff Engineer?

If the only career path was to become a manager (like in the company depicted on the left in Figure 1-1), many engineers would be faced with a stark and difficult choice: stay in an engineering role and keep growing in their craft or move to management and grow in their careers instead.

So it's good that many companies now offer a "technical" or "individual contributor" track, allowing career progression in parallel to manager roles. The ladder on the right in Figure 1-1 shows an example.

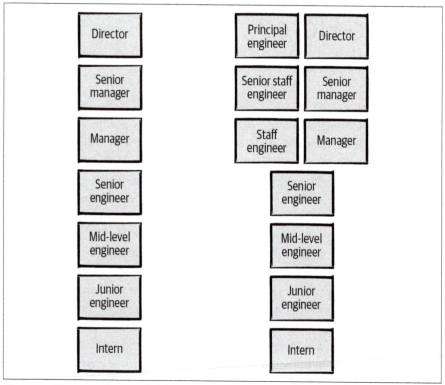

Figure 1-1. Two example career ladders, one with multiple paths.

Job ladders vary from company to company, enough that it's given rise to a website, *levels.fyi*, that compares technical track ladders across companies.[1] The number of rungs on these ladders varies, as do the names of each rung. You may even see the same names in a different order.[2] But, very often, the word *senior* is

---

1 I also recommend *progression.fyi*, which has an extensive collection of ladders published by various tech companies.

2 One company I heard about used the levels "senior," "staff," and "principal," in that order of seniority, but got acquired by another company that used "senior," "principal," and "staff." Chaos. The acquiring company changed all "staff" to "principal" and all "principal" to "staff," and no one was happy. Both staffs and principals saw the change as a demotion. Titles matter!

used. Marco Rogers, a director of engineering who has created career ladders at two companies, has described (*https://oreil.ly/MpwsJ*) the *senior* level as the "anchor" level for a career ladder. As Rogers says, "The levels below are for people to grow their autonomy; the levels above increase impact and responsibility."

Senior is sometimes seen as the "tenure" level: you don't need to go further.[3] But if you do, you enter the "technical leadership" levels. The first rung above senior is often called "staff engineer," and that's the name I'll use throughout this book.

In the dual-track job ladder from Figure 1-1, a senior engineer can choose to build the skills to get promoted to either a manager or a staff engineer role. Once they've been promoted, a role change from staff engineer to manager, or vice versa, would be considered a sideways move, not a further promotion. A senior staff engineer would have the same seniority as a senior manager, a principal engineer would equate to a director, and so on; those levels might continue even higher in the company's career ladders. (To represent all of the roles above senior, I'm going to use *staff+*, an expression coined by Will Larson in his book *Staff Engineer.*)

A Note About Titles

I've occasionally heard people insist that job titles and leveling shouldn't (or don't) matter. People who make this claim tend to say reasonable things about their company being an egalitarian meritocracy that is wary of the dangers of hierarchy. "We're a bottom-up culture and all ideas are treated with respect," they say, and that's an admirable goal: being early in your career should never mean your ideas are dismissed.

But titles do matter. The Medium engineering team wrote a blog post (*https://oreil.ly/oUkHe*) that lays out three reasons titles are necessary: "Helping people understand that they are progressing, vesting authority in those people who might not automatically receive it, and communicating an expected competency level to the outside world."

While the first reason is intrinsic and, perhaps, not a motivation for everyone, the other two describe the effect that a title has on other

---

3 I like my friend Tiarnán de Burca's definition of senior engineer: the level at which someone can stop advancing and continue their current level of productivity, capability, and output for the rest of their career and still be "regretted attrition" if they leave.

people. Whether a company claims to be flat and egalitarian or not, there will always be those who react differently to people of different levels, and most of us are at least a little status-conscious. As Dr. Kipp Krukowski, clinical professor of entrepreneurship at Colorado State University, says in his 2017 paper, "The Effects of Employee Job Titles on Respect Granted by Customers" (*https://oreil.ly/zD3kp*), "Job titles act as symbols and companies use them to signal qualities of their workers to individuals both inside and outside of the firm."

We make implicit judgements and assumptions about people all the time. Unless we've invested a lot of time and energy in becoming aware of our implicit biases, it's likely that these assumptions will be influenced by stereotypes. A 2015 survey (*https://oreil.ly/snmmY*), for example, found that around half of the 557 Black and Latina professional women in STEM surveyed had been mistaken for janitors or administrative staff.

When a software engineer walks into a meeting with people they don't know, similar implicit biases come into play. White and Asian male software engineers are often assumed to be more senior, more "technical," and better at coding, whether they graduated yesterday or have been doing the job for decades. Women, especially women of color, are assumed to be more junior and less qualified. They have to work harder in the meeting to be assumed competent.

As that Medium engineering article said, a job title vests authority in people who might not automatically receive it, and communicates their expected competency level. By anchoring expectations, it saves them the time and energy they would otherwise have to spend proving themselves again and again. It gives them some hours back in their week.

The title you have now also influences the job you'll have next. Like many folks in our industry, I get daily emails from recruiters on LinkedIn. *Exactly three times* in my life I've had a cold-call recruiting email that invited me to interview for a more senior job title than the one I already had. All others have suggested a role at exactly the level that I was already at, or a more junior one.

So that's *what* the job looks like on a ladder. But let's look at *why* the technical leadership levels exist. I talked in the introduction about the three pillars of the technical track: big-picture thinking, project execution, and leveling up. Why do we need *engineers* to have those skills? Why do we need staff engineers at all?

WHY DO WE NEED ENGINEERS WHO CAN SEE THE BIG PICTURE?

Any engineering organization is constantly making decisions: choosing technology, deciding what to build, investing in a system or deprecating it. Some of these decisions have clear owners and predictable consequences. Others are foundational architectural choices that will affect every other system, and no one can claim to know exactly how they'll play out.

Good decisions need *context*. Experienced engineers know that the answer to most technology choices is "it depends." Knowing the pros and cons of a particular technology isn't enough—you need to know the local details too. What are you trying to do? How much time, money, and patience do you have? What's your risk tolerance? What does the business need? That's the context of the decision.

Gathering context takes time and effort. Individual teams tend to optimize for their own interests; an engineer on a single team is likely to be laser-focused on achieving that team's goals. But often decisions that seem to belong to one team have consequences that extend far beyond that team's boundaries. The *local maximum*, the best decision for a single group, might not be anything like the best decision when you take a broader view.

Figure 1-2 shows an example where a team is choosing between two pieces of software, A and B. Both have the necessary features, but A is significantly easier to set up: it just works. B is a little more difficult: it will take a couple of sprints of wrangling to get it working, and nobody's enthusiastic about waiting that long.

From the team's point of view, A is a clear winner. Why would they choose anything else? But other teams would much prefer that they choose B. It turns out that A will make ongoing work for the legal and security teams, and its authentication needs mean the IT and platform teams will have to treat it as a special case forever. By choosing A, the local maximum, the team is unknowingly choosing a solution that's a much bigger time investment for the company overall. B is only slightly worse for the team, but much better overall. Those extra two sprints will pay for themselves within a quarter, but this fact is only obvious when the team has someone who can look through a wider lens.

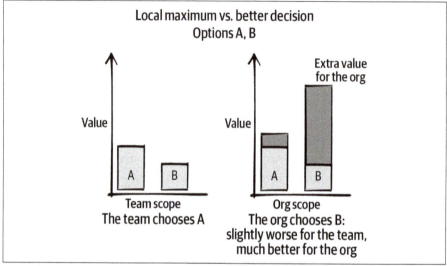

Figure 1-2. Local maximum versus better decision.

To avoid local maxima, teams need decision makers (or at least decision influencers) who can take an *outsider view*—who can consider the goals of multiple teams at once and choose a path that's best for the whole organization or the whole business. Chapter 2 will cover zooming out and looking at the bigger picture.

Just as important as seeing the big picture of the situation *now* is being able to anticipate how your decisions will play out in future. A year from now, what will you regret? In three years, what will you wish you'd started doing now? To travel in the same direction, groups need to agree on technical strategies: which technologies to invest in, which platforms to standardize on, and so on. These huge decisions can end up being subtle, and they're often controversial, so essential to making the decision is being able to share context and help others make sense of it. Chapter 3 is all about choosing a direction as a group.

So, if you want to make broad, forward-looking decisions, you need people who can see the big picture. But why can't that be managers? And why can't the chief technology officer (CTO) just know all of the "business things," translate them into technical outcomes, and pass on what matters?

On some teams, they can. For a small team, a manager can often function as the most experienced technologist, owning major decisions and technical direction. In a small company, a CTO can stay deeply involved in the gory details of every decision. These companies probably don't need staff engineers. But management authority can overshadow technical judgment: reports may feel uncomfortable arguing with a manager's technical decisions even when there's a better solution available. And managing other humans is itself a full-time job. Someone who's investing in being a good people manager will have less time available to stay up to date with technical developments, and anyone who is managing to stay deeply "in the weeds" will be less able to meet the needs of their reports. In the short term that can be OK: some teams don't need a lot of attention to continue on a successful path. But when there's tension between the needs of the team and the needs of the technical strategy, a manager has to choose where to focus. Either the team's members or its technical direction get neglected.

That's one reason that many organizations create separate paths for technical leadership and people leadership. If you have more than a few engineers, it's inefficient—not to mention disempowering—if every decision needs to end up on the desk of the CTO or a senior manager. You get better outcomes and designs if experienced engineers have the time to go deep and build the context and the authority to set the right technical direction.

That doesn't mean engineers set technical direction alone. Managers, as the people responsible for assigning headcount to technical initiatives, need to be part of major technical decisions. I'll talk about maintaining alignment between engineers and managers later in this chapter, and again when we're talking strategy in Chapter 3.

What About Architects?

In some companies, "architect" is a rung on the technical track of the job ladder. In others, architects are abstract system designers who have their own career path, distinct from that of the engineers who implement the systems. In this book I'm going to consider software design and architecture to be part of the role of a staff+ engineer, but be aware that this is not universally true in our industry.

WHY DO WE NEED ENGINEERS WHO LEAD PROJECTS THAT CROSS MULTIPLE TEAMS?

In an ideal world, the teams in an organization should interlock like jigsaw puzzle pieces, covering all aspects of any project that's underway. In this same ideal world, though, everyone's working on a beautiful new green-field project with no prior constraints or legacy systems to work around, and each team is wholly dedicated to that project. Team boundaries are clear and uncontentious. In fact, we're starting out with what the Thoughtworks tech consultants have dubbed an Inverse Conway Maneuver (*https://oreil.ly/HdKyK*): a set of teams that correspond exactly with the components of the desired architecture. The difficult parts of this utopian project are difficult only because they involve deep, fascinating research and invention, and their owners are eager for the technical challenge and professional glory of solving them.

I want to work on that project, don't you? Unfortunately, reality is somewhat different. It's almost certain that the teams involved in any cross-team project already existed before the project was conceived and are working on other things, maybe even things that they consider more important. They'll discover unexpected dependencies midway through the project. Their team boundaries have overlaps and gaps that leak into the architecture. And the murky and difficult parts of the project are not fascinating algorithmic research problems: they involve spelunking through legacy code, negotiating with busy teams that don't want to change anything, and divining the intentions of engineers who left years ago.[4] Even understanding what needs to change can be a complex problem, and not all of the work can be known at the start. If you look closely at the design documentation, you might find that it postpones or hand-waves the key decisions that need the most alignment.

That's a more realistic project description. No matter how carefully you overlay teams onto a huge project, some responsibilities end up not being owned by anyone, and others are claimed by two teams. Information fails to flow or gets mangled in translation and causes conflict. Teams make excellent *local maximum* decisions and software projects get stuck.

One way to keep a project moving is to have someone who feels ownership for the whole thing, rather than any of its individual parts. Even before the project kicks off, that person can scope out the work and build a proposal. Once the project is underway, they're likely to be the author or coauthor of the high-level system design and a main point of contact for it. They maintain a high engineering standard, using their experience to anticipate risks and ask hard questions. They also spend time informally mentoring or coaching—or just

---

4 What were they *thinking*? Was this really what they intended to do? Of course, future teams will ask the same of us.

setting a good example for—the leads of individual parts of the project. When the project gets stuck, they have enough perspective to track down the causes and unblock it (more on that in Chapter 6). Outside the project, they're telling the story of what's happening and why, selling the vision to the rest of the company, and explaining what the work will make possible and how the new project affects everyone.

Why can't technical program managers (TPMs) do this consensus-building and communication? There is definitely some overlap in responsibilities. Ultimately, though, TPMs are responsible for delivery, not design, and not engineering quality. TPMs make sure the project gets *done on time*, but staff engineers make sure it's done with high engineering standards. Staff engineers are responsible for ensuring the resulting systems are robust and fit well with the technology landscape of the company. They are cautious about technical debt and wary of anything that will be a trap for future maintainers of those systems. It would be unusual for TPMs to write technical designs or set project standards for testing or code review, and no one expects them to do a deep dive through the guts of a legacy system to make a call on which teams will need to integrate with it. When a staff engineer and TPM work well together on a big project, they can be a dream team.

WHY DO WE NEED ENGINEERS WHO ARE A GOOD INFLUENCE?

Software matters. The software systems we build can affect people's well-being and income: Wikipedia's list of software bugs (*https://oreil.ly/eNIXO*) makes for good, if sobering, reading. We've learned from plane crashes (*https://oreil.ly/iJgF2*), ambulance system failures (*https://oreil.ly/s9GQf*), and malfunctioning medical equipment (*https://oreil.ly/fr7Dj*) that software bugs and outages can kill people, and it would be naive to assume there won't be more and bigger software-related tragedies coming in our future.[5] We need to take software seriously.

Even when the stakes are lower, we're still making software for a reason. With a few R&D-ish exceptions, engineering organizations usually don't exist just for the sake of building more technology. They're setting out to solve an

---

5 Hillel Wayne's essay "We Are Not Special" (*https://oreil.ly/WKOTK*) points out that a lot of engineering solutions that used to involve carefully tuning physical equipment are now done with a "software kludge" instead. I'm genuinely always surprised we've had so few major fatal accidents from software so far. I wouldn't like to depend on us staying lucky.

actual business problem or to create something that people will want to use. And they'd like to achieve that with some acceptable level of quality, an efficient use of resources, and a minimum of chaos.

Of course, quality, efficiency, and order are far from guaranteed, particularly when there are deadlines involved. When doing it "right" means going slower, teams that are eager to ship may skip testing, cut corners, or rubber-stamp code reviews. And creating good software isn't easy or intuitive. Teams need senior people who have honed their skills, who have seen what succeeds and what fails, and who will take responsibility for creating software that works.

We learn from every project, but each of us has only a finite number of experiences to reflect on. That means that we need to learn from *each other's* mistakes and successes, too. Less experienced team members might never have seen good software being made, or might see producing code as the only important skill in software engineering. More seasoned engineers can have huge impact by conducting code and design reviews, providing architectural best practices, and creating the kinds of tooling that make everyone faster and safer.

Staff engineers are role models. Managers may be responsible for setting culture on their teams, enforcing good behavior, and ensuring standards are met. But engineering norms are set by the behavior of the most respected engineers on the project. No matter what the standards say, if the most senior engineers don't write tests, you'll never convince everyone else to do it. These norms go beyond technical influence: they're cultural, too. When senior people vocally celebrate other people's work, treat each other with respect, and ask clarifying questions, it's easier for everyone else to do that too. When early-career engineers respect someone as the kind of engineer they want to "grow up" to be, that's a powerful motivator to act like they do. (Chapter 7 will explore leveling up your organization by being a role model.)

Maybe now you're convinced that engineers should do this big-picture, big-project, good-influence stuff, but here's the problem: they can't do it on top of the coding workload of a senior engineer. Any hour you're writing strategy, reviewing project designs, or setting standards, you're not coding, architecting new systems, or doing a lot of the work a software engineer might be evaluated on. If a company's most senior engineers just write code all day, the codebase will see the benefit of their skills, but the company will miss out on the things that only they can do. This kind of technical leadership needs to be part of the job description of the person doing it. It isn't a distraction from the job: it *is* the job.

Enough Philosophy. What's My Job?

The details of a staff engineering role will vary. However, there are some attributes of the job that I think are fairly consistent. I'll lay them out here, and the rest of the book will take them as axiomatic.

YOU'RE NOT A MANAGER, BUT YOU ARE A LEADER

First things first: staff engineering is a *leadership* role. A staff engineer often has the same seniority as a line manager. A principal engineer often has the seniority of a director. As a staff+ engineer, you're the counterpart of a manager at the same level, and you're expected to be as much "the grown-up in the room" as they are. You may even find that you're more senior and more experienced than some of the managers in your organization. Whenever there's a feeling of "someone should do something here," there's a reasonable chance that the someone is you.

Do you *have* to be a leader? Midlevel engineers sometimes ask me if they *really* need to get good at "that squishy human stuff" to go further. Aren't technical skills enough? If you're the sort of person who got into software engineering because you wanted to do technical work and don't love talking to other humans, it can feel unfair that your vocation runs into this wall. But if you want to keep growing, being deep in the technology can only take you so far. Accomplishing larger things means working with larger groups of people—and that needs a wider set of skills.

As your compensation increases and your time becomes more and more expensive, the work you do is expected to be more valuable and have a greater impact. Your technical judgment will need to include the reality of the business and whether any given project is worth doing at all. As you increase in seniority, you'll take on bigger projects, projects that can't succeed without collaboration, communication, and alignment; your brilliant solutions are just going to cause you frustration if you can't convince the other people on the team that yours is the right path to take. And whether you want to or not, you'll be a role model: other engineers will look to those with the big job titles to understand how to behave. So, no: you can't avoid being a leader.

Staff engineers lead differently than managers, though. A staff engineer usually doesn't have direct reports. While they're involved and invested in growing the technical skills of the engineers around them, they're not responsible for managing anyone's performance or approving vacation or expenses. They can't fire or promote—though local team managers should value their opinions about other team members' skills and output. Their impact happens in other ways.

Leadership comes in lots of forms that you might not immediately recognize as such. It can come from designing "happy path" solutions that protect other engineers from common mistakes. It can come from reviewing other engineers' code and designs in a way that improves their confidence and skills, or from highlighting that a design proposal doesn't meet a genuine business need. Teaching is a form of leadership. Quietly raising everyone's game is leadership. Setting technical direction is leadership. Finally, there's having the reputation as a stellar technologist that can inspire other people to buy into your plans just because they trust you. If that sounds like you, then guess what? You're a leader.

Yes, You Can Be an Introvert. No, You Can't Be a Jerk.

The idea of "being a leader" can be a little intimidating for many people. Don't worry: not all staff and principal engineers need to be "people people." Staff engineering has plenty of room for introverts—and even the quietest engineers can set a strong technical direction through their judgment and good influence. You don't have to love being around people to be a good leader. You do have to be a role model, though, and you have to treat people well.

Many of us even have stories of "that one engineer" who got shuffled into a corner because they were too difficult for anybody to deal with. The tech culture of the 1980s and 1990s, exemplified by discussions on Usenet and the like, reveled in the popular image (*https://en.wikipe dia.org/wiki/Bastard_Operator_From_Hell*) of the difficult, unpleasant software engineer, whose colleagues not only tolerated their behavior but made weird technical decisions just to avoid dealing with them. Today, however, an engineer like this is a liability. No matter what their output is, it's hard to imagine how anyone could be worth the reduced output and growth of other engineers and the projects that fail when that engineer won't collaborate across teams. Choosing these people as role models can mess up whole organizations.

If you suspect your colleagues will think this sidebar is about you, check out Kind Engineering (*https://kind.engineering*), where Evan Smith, SRE manager at Squarespace, gives concrete advice on how to be an actively kind coworker. You'll be surprised at how quickly you can turn around a reputation for being difficult to work with.

YOU'RE IN A "TECHNICAL" ROLE

Staff engineering is a leadership role, but it's also a deeply specialized one. It needs technical background and the kinds of skills and instincts that come from engineering experience. To be a good influence, you need to have high standards for what excellent engineering looks like and model them when you build something. Your reviews of code or designs should be instructive for your colleagues and should make your codebase or architecture better. When you're making technical decisions, you need to understand the trade-offs and help other people understand them too. You need to be able to dive into the details where necessary, ask the right questions, and understand the answers. When arguing for a particular course of action, or a particular change in technical culture, you need to know what you're talking about. So you have to have a solid foundation of technical skills.

This doesn't necessarily mean you'll write a lot of code. At this level, your goal is to solve problems efficiently, and programming will often not be the best use of your time. It may make more sense for you to take on the design or leadership work that only you can do and let others handle the programming. Staff engineers often take on ambiguous, messy, difficult problems and do just enough work on them to make them manageable by someone else. Once the problem is tractable, it becomes a growth opportunity for less experienced engineers (sometimes with support from the staff engineer).

For some staff engineers, deep diving through codebases will remain the most efficient tool to solve many problems. For others, writing documents might get better results, or becoming a master of data analysis, or having a terrifying number of one-on-one meetings. What matters is *that* the problems get solved, not *how*.[6]

---

6 This is why I'm not a fan of giving experienced staff engineers coding interviews. If you've made it to this level, either you can code well or you've learned to solve technical problems using your other muscles. The outcomes are what matters.

YOU AIM TO BE AUTONOMOUS

When you started out as an engineer, your manager probably told you what to work on and how to approach it. At senior level, maybe your manager advised you on which problems were important to solve, and left it to you to figure out what to do about it. At staff+ levels, your manager should be bringing you information and sharing context, but *you* should be telling *them* what's important just as much as the other way around. As Sabrina Leandro, principal engineer at Intercom, asks (*https://oreil.ly/FOI1L*), "So you know you're supposed to be working on things that are impactful and valuable. But where do you find this magic backlog of high-impact work that you should be doing?" Her answer: "You create it!"

As a senior person in the organization, it's likely that you'll be pulled in many directions. It's up to you to defend and structure your time. There are a finite number of hours in the week (see Chapter 4). You get to choose how to spend them. If someone asks you to work on something, you'll bring your expertise to the decision. You'll weigh the priority, the time commitment, and the benefits—including the relationship you want to maintain with the person who asked you for help—and you'll make your own call. If your CEO or other local authority figure tells you they need something done, you'll give that appropriate weight. But autonomy demands responsibility. If the thing they asked you to work on turns out to be harmful, you have a responsibility to speak up. Don't silently let a disaster unfold. (Of course, if you want to be listened to, you'll have to have built up a reputation for being trustworthy and correct.)

YOU SET TECHNICAL DIRECTION

As a technical leader, part of a staff engineer's role is to make sure the organization has a good technical direction. Underlying the product or service your organization provides is a host of technical decisions: your architecture, your storage systems, the tools and frameworks you use, and so on. Whether these decisions are made at a team level or across multiple teams or whole organizations, part of your job is to make sure that they get made, that they get made well, and that they get written down. The job is not to come up with all (or even necessarily any!) of the aspects of the technical direction, but to ensure there is an agreed-upon, well-understood solution that solves the problems it sets out to solve.

YOU COMMUNICATE OFTEN AND WELL

The more senior you become, the more you will rely on strong communication skills. Almost everything you do will involve conveying information from your brain to other people's brains and vice versa. The better you are at being understood, the easier your job will be.

Understanding Your Role

Those axioms should help you to start defining your role, but you'll notice that they leave out a lot of implementation details! The truth is that the day-to-day work of one staff engineer might look very different from that of another. The realities of your role will depend on the size and needs of your company or organization, and will also be influenced by your personal work style and preferences.

This variation means that it can be hard to compare your work to that of staff engineers around you or in other companies. So in this section, we're going to unpack some of the role's more variable attributes.

Let's start with reporting chains.

WHERE IN THE ORGANIZATION DO YOU SIT?

Our industry hasn't settled on any standard model for how staff+ engineers report into the rest of the engineering organization. Some companies have their most senior engineers report to a chief architect or the office of the CTO; others assign them to directors of various organizations, to managers at various levels, or to a mix of all of the above. There's no one right answer here, but there can be a lot of wrong answers, depending on what you're trying to achieve.

Reporting chains (see the example in Figure 1-3) will affect the level of support you receive, the information you're privy to, and, in many cases, how you're perceived by colleagues outside your group.

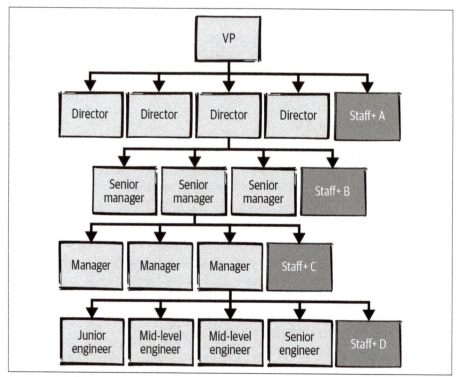

Figure 1-3. Staff+ engineers reporting in at different levels of the org hierarchy. Even if these engineers are all at the same level of seniority, A will find it much easier to have organizational context and to be in director-level conversations than D will.

Reporting "high"

Reporting "high" in the org chart, such as to a director or VP, will give you a broad perspective. The information you get will be high-level and impactful, and so will the problems you're asked to solve. If you're reporting to a very competent senior person, watching them make decisions, run meetings, or navigate a crisis can be a unique and valuable learning experience.

That said, you'll probably get a lot less of your manager's time than you would if you had a local manager. Your manager might have less visibility into your work and therefore might not be able to advocate for you or help you grow. An engineer working closely with a single team but reporting to a director may feel disconnected from the rest of the team or might pull the director's attention into local disagreements that should have been solved at the team level.

If you find that your manager isn't available, doesn't have time to understand the work that you do, or gets pulled into low-level technical decisions that aren't a good use of their time, consider that you might be happier with a manager whose focus is more aligned with yours.

Reporting "low"

Reporting to a manager lower in the org chart brings its own set of advantages and disadvantages. Chances are that you'll get more focused attention from your manager, and you'll be more likely to have an advocate. If you prefer to focus on a single technical area, you might benefit from working with a manager who is close to that area.

But an engineer assigned to a single team may find it hard to influence the whole organization. Like it or not, humans pay attention to status and hierarchies —and reporting chains. You're likely to have much less influence if you're reporting to a line manager. The information you get is also prone to be more filtered and centered on the problems of that specific team. If your manager doesn't have access to some piece of information, you almost certainly won't either.

Reporting to a line manager may also mean that you're reporting to someone less experienced than you are. That's not inherently a problem, but you may have less to learn from your manager, and they might not be helpful for career development: chances are that they won't know how to help you. All of this may be fine if you're getting some of your management needs met elsewhere.[7] In particular, if you're reporting to someone low in the org hierarchy, make sure to have *skip-level* meetings with your manager's manager.[8] Find ways to stay connected to your organization's goals.

If you and your manager have different ideas about how you can be most effective, that can cause tension. You can end up with a case of the local maximum issues I mentioned earlier, where your manager wants you to work on the most important concern of the *team*, when there are far bigger problems inside the *organization* that need you more. It's harder for a technical or prioritization debate to happen on a truly level playing field when one person is responsible for the other's performance rating and compensation. If you find that these arguments are happening a lot, you might want to advocate to report to a level higher.

---

7 I recommend Lara Hogan's article (*https://oreil.ly/wY9Mp*) on building a "manager Voltron."

8 If skip-level meetings aren't common in your company, you may need to be clear that you're not looking to undermine or "report" on your manager; you want to understand the wider group's priorities and make the connections that can help you have the most impact. Ideally, your manager will understand the value of skip-level meetings and help you set them up.

WHAT'S YOUR SCOPE?

Your reporting chain will likely affect your *scope*: the domain, team, or teams that you pay close attention to and have some responsibility for, even if you don't hold any formal leadership role in this domain.

Inside your scope, you should have some influence on short-term and long-term goals. You should be aware of the major decisions being made. You should have opinions about changes and represent people who don't have the leverage to prevent poor technical decisions that affect them. You should be thinking about how to cultivate and develop the next generation of senior and staff engineers, and should notice and suggest projects and opportunities that would help them grow.

In some cases, your manager might expect you to devote the majority of your skills and energy to solving problems that fall within their domain. In other cases, a team may just be a home base as you spend some portion of your time on fires or opportunities elsewhere in the org. If you report to a director, there may be an implicit assumption that you operate at a high level and tie together the work of everything that's happening in the org, or you might be explicitly allocated to some subset of the director's teams or technology areas. Be clear about which it is.

Be prepared to ignore your scope when there's a crisis: there is no such thing as "not my job" during an outage, for example. You should also have a level of comfort with stepping outside your day-to-day experience, leading when needed, learning what you need to learn, and fixing what you need to fix. Part of the value of a staff engineer is that you *don't* stay in your lane.

Nonetheless, I recommend that you get very clear on what your scope is, even if it's temporary and subject to change.

A scope too broad

If your scope is too broad (or undefined), there are a few possible failure modes.

Lack of impact

> If *anything* can be your problem, then it's easy for *everything* to become your problem, particularly if you're in an organization with fewer senior people than it needs. There will always be another side quest: in fact, it's all

too easy to create a role that's *entirely* side quests, with no real goal at all.[9] Beware of spreading yourself too thin. You can end up without a narrative to your work that makes you (and whoever hired you) feel like you achieved something.

Becoming a bottleneck
When there's a senior person who is seen to do everything, the convention can become that they *need* to be in the room for every decision. Rather than speeding up your organization, you end up slowing them down because they can't manage without you.

Decision fatigue
If you escape the trap of trying to do everything, you'll have the constant cost of deciding *which* things to do. I'll talk in Chapter 4 about choosing your work.

Missing relationships
If you're working with a very broad set of teams, it's harder to have enough regular contact to build the sorts of friendly relationships that make it easier to get things done (and that make work enjoyable!). Other engineers also lose out: they don't get the sort of mentorship and support that comes from having a "local" staff engineer involved in their work.

It's hard to operate in a workplace where you can do literally anything. Better to choose an area, build influence, and have some successes there. Devote your time to solving some problems entirely. Then, if you're ready to, move on to a different area.

A scope too narrow

Beware, too, of scoping yourself too narrowly. A common example is when a staff engineer is part of a single team, reporting to a line manager. Managers might really like this—they get a very experienced engineer who can do a large percentage of the design and technical planning, and perhaps serve as a technical leader or team lead for a project. Some engineers will love this too: it means you get to

---

9 A side quest (*https://oreil.ly/LDRd5*) is a part of a video game that doesn't have anything to do with the main mission, but that you can optionally do for coins or experience points or just for fun. Picture lots of, "Well, I was about to fight my way into the heavily guarded fortress to defeat the demon that's been terrorizing this land, but sure, I can go find your cat first."

go really deep on the team's technologies and problems and understand all of the nuances. But watch out for the risks of a scope that's too narrow:

Lack of impact

It's possible to spend all of your time on something that doesn't *need* the expertise and focus of a staff engineer. If you choose to go really deep on a single team or technology, it should be a core component, a mission-critical team, or something else that's very important to the company.

Opportunity cost

Staff engineers' skills are usually in high demand. If you're assigned to a single team, you may not be top of mind for solving a problem elsewhere in the org, or your manager may be unwilling to let you go.

Overshadowing other engineers

A narrow scope can mean that there's not enough work to keep you busy, and that you may overshadow less experienced people and take learning opportunities away from them. If you always have time to answer all of the questions and take on all of the tricky problems, nobody else gets experience in doing that.

Overengineering

An engineer who's not busy can be inclined to make work for themselves. When you see a vastly overengineered solution to a straightforward problem, that's often the work of a staff engineer who should have been assigned to a harder problem.

Some technical domains and projects are deep enough that an engineer can spend their whole career there and never run out of opportunities. Just be very clear about whether you're in one of those spaces.

WHAT SHAPE IS YOUR ROLE?

So long as it's generally agreed that your work is impactful, you should have a lot of flexibility around how you do it. That includes a certain amount of defining what your job is. Here are a few questions to ask yourself:

Do you approach things depth-first or breadth-first?

Do you prefer to focus narrowly on a single problem or technology area? Or are you more inclined to go broad across multiple teams or technologies, focusing on a single problem only when it can't be solved without you? Being depth-first or breadth-first is very much about your personality and work style.

There's no wrong answer here, but you'll have an easier and more enjoyable time if your preference here is lined up with your scope. For instance, if you want to influence the technical direction of your org or business, you'll find yourself gravitating toward opportunities to take a broader view. You'll need to be in the rooms where the decisions are happening and tackle problems that affect many teams. If you're trying to do that while assigned to a single deep architectural problem, no one wins. On the other hand, if you're aiming to become an industry expert in a particular technical domain, you'll need to be able to narrow your focus and spend most of your time in that one area.

Which of the "four disciplines" do you gravitate toward?

Yonatan Zunger, distinguished engineer at Twitter, describes the four disciplines (*https://oreil.ly/3S9HE*) that are needed in any job in the world:

Core technical skills
> Coding, litigation, producing content, cooking—whatever a typical practitioner of the role works on

Product management
> Figuring out what needs to be done and why, and maintaining a narrative about that work

Project management
> The practicalities of achieving the goal, removing chaos, tracking the tasks, noticing what's blocked, and making sure it gets unblocked

People management
> Turning a group of people into a team, building their skills and careers, mentoring, and dealing with their problems

Zunger notes that the higher your level, the less your mix of these skills corresponds with your job title: "The more senior you get, the more this becomes true, the more and more there is an expectation that you can shift across each of these four kinds of jobs easily and fluidly, and function in all rooms."

Every team and every project needs all four of these skills. As a staff engineer, you'll use all of them. You don't need to be amazing at all of them, though. We all have different aptitudes and enjoy or avoid different kinds of work. Maybe it's obvious to you which ones you enjoy and which you hope to never need. If you're not sure, Zunger suggests discussing each one with a friend and having them watch your emotional response and energy while you talk about it. If there's one that you *really* hate, make sure you're working with someone who's eager to do that aspect of the work. Whether you're breadth-first or depth-first, you'll find it hard to continue to grow with *only* the core technical skills.

The Hyperspecialist Career Path

There are a few rare cases where a strong senior engineer in a *very business-critical domain* can be successful without planning ahead or influencing people around them. Zunger calls this the "hyperspecialist" role, but notes that "over time your influence will wane. There are actually very few jobs at senior levels that are purely hyperspecialists. It's not a thing people tend to need." Pat Kua calls this path "the true individual contributor track" (*https://oreil.ly/9IFOB*), noting that it still needs excellent communication and collaboration skills. Depending on the company, the "hyperspecialist" path may be considered a staff engineer role or be entirely separate.

How much do you want (or need) to code?

For "coding" here, feel free to swap in the core technical work of your career so far. This set of skills probably got you to where you are today, and it can be uncomfortable to feel that you're getting rusty or out of date. Some staff engineers find that they end up reading or reviewing a lot of code but not writing much at all. Others are core contributors to projects, coding every day. A third group *finds* reasons to code, taking on noncritical projects that will be interesting or educational but won't delay the project.

If you're going to feel antsy unless you're in code every day, make sure you're not taking on a broad architectural or influence-based role where you just won't have time. Or at least have a plan for how you're going to scratch that itch, so you'll be able to resist jumping on coding tasks and leaving the bigger problems to fend for themselves.

How's your delayed gratification?

Coding has comfortingly fast feedback cycles: every successful compile or test run tells you how things are going. It's like a tiny performance review every day!

It can be disheartening to move toward work that doesn't have any built-in feedback loops to tell you whether you're on the right path.

On a long-term or cross-organizational project, or with strategy or culture change, it can be months—or even longer—before you have a strong signal about whether what you're doing is working. If you're going to be anxious and stressed out on a project with longer feedback cycles, ask a manager who you trust to tell you, regularly and honestly, how things are going. If you need that and don't have it, consider projects that pay off on a shorter timescale.

Are you keeping one foot on the manager track?

Although most staff engineers don't have direct reports, some do. A tech lead manager (TLM), sometimes called a team lead, is a kind of hybrid role where the staff engineer is the technical leader for a team and also manages that team. It's a famously (*https://oreil.ly/uRrBq*) difficult (*https://oreil.ly/8eFBM*) gig (*https://oreil.ly/8S4vR*). It can be challenging to be responsible for both the humans and the technical outcomes without feeling like you're failing at one or the other. It's also difficult to find time to invest in building skills on either side, and I've heard TLM folks lament a loss of career progression as a result.

Some people take a management role for a couple of years, then a staff engineer role, going back and forth every so often to keep their skills sharp on both sides.[10] We'll look more at this "pendulum" and at TLM roles in Chapter 9.

Do any of these archetypes fit you?

In his article "Staff Archetypes" (*https://oreil.ly/cYVGl*), Will Larson describes four distinct patterns he's seen staff engineering roles take. You can use these archetypes as you define the kind of role you have, or would like to have:

Tech leads
> Partner with managers to guide the execution of one or more teams.

Architects
> Responsible for technical direction and quality across a critical area.

Solvers
> Wade into one difficult problem at a time.

---

10 Charity Majors's "The Engineer/Manager Pendulum" (*https://oreil.ly/aV16i*) is an excellent article on this topic.

Right hands
Add leadership bandwidth to an organization.

If you don't see yourself in any of those archetypes, or your role crosses more than one of them, that's OK! These archetypes are not intended to be prescriptive; they give us concepts to use in articulating how we prefer to work.

WHAT'S YOUR PRIMARY FOCUS?

So we've discussed your scope and your reporting chain: the rough boundaries of the part of the organization you're operating inside, and where in the organization you sit. We've also looked at your aptitudes: how you like to work and what kinds of skills you're drawn to. But even if you understand all of that and have a clear picture of the shape of your role, there's one question left: what are you going to work on?

As you grow in influence, you'll find that more and more people want you to *care* about things. Someone's putting together a best practices document for how your organization does code review, and they want your opinion. Your group is doing a hiring push and needs help deciding what to interview for. There's a deprecation that would be making more progress if it had a staff engineer drumming up senior sponsorship. And that's just Monday morning. What do you do?

In some cases, your manager or someone they report to will have strong opinions about where you should focus, or will even have hired you specifically to solve a particular problem. Most of the time, though, you'll have some autonomy in deciding what's most important. Every time you choose what to work on, you're also choosing what *not to do*, so be deliberate and thoughtful about what you take on.

What's important?

Early in your career, if you do a great job on something that turns out to be unnecessary, you've still done a great job. At the staff engineer level, though, everything you do has a high opportunity cost, so your work needs to be *important.*

Let's unpack that for a moment. "Your work needs to be important" doesn't mean you should only work on the fanciest, most glamorous technologies and VP-sponsored initiatives. The work that's most important will often be the work that nobody else sees. It might be a struggle to even articulate the need for it, because your teams don't have good mental models for it yet. It might involve gathering data that doesn't exist, or spelunking through dusty code or documents that haven't been touched in a decade. There are any number of other grungy tasks that just need to get done. Meaningful work comes in many forms.

Know why the problem you're working on is strategically important—and if it's not, do something else.

What needs you?

There's a similar situation when a senior person devotes themself to the sort of coding project that any midlevel engineer could have taken on: you're going to do a stellar job on it, but chances are there's a senior-sized problem available that the midlevel engineer wouldn't be able to tackle. To use an idiom my kid dropped profoundly one day, "You don't plant grass in your only barrel."

Be wary of choosing a project that already has a lot of senior people on it. Scope out who else is working on the problem and whether they seem likely to succeed at solving it. Some projects may even be slowed by an extra leader joining.[11] In general, if there are more people being the wise voice of reason than there are people actually typing code (or whatever your project's equivalent is), don't butt in. Try to choose a problem that actually needs you and that will benefit from your attention. Chapter 4 will give you some tools for deciding which projects to take on.

Aligning on Scope, Shape, and Primary Focus

By now, you should have a pretty clear picture of what the scope of your role is, how it's shaped, and what you're working on right now. But are you certain that your picture matches everyone else's? Your manager's and colleagues' expectations may differ wildly from yours on what a staff engineer is, what authority you have to make decisions, and myriad other big questions. If you're joining a company as a staff engineer, it's best to get all of this straightened out up front.

A technique I learned from my friend Cian Synnott is to write out my understanding of my job and share it with my manager. It can feel a little intimidating to answer the question "What do you do here?" What if other people think what you do is useless, or think you don't do it well? But writing it out removes the ambiguity, and you'll find out early if your mental model of the role is the same as everyone else's. Better now than at performance review time.

---

11 You'll hear Brooks's Law quoted: "Adding manpower to a late software project makes it later." While Brooks himself called this "an outrageous simplification" (*https://oreil.ly/WlruQ*), there's truth to it. See *The Mythical Man-Month* by Fred Brooks (Addison-Wesley).

Here's what such a role description might look like for Ali, a breadth-first architect-archetype staff engineer, who is assisting with (but not leading) a large cross-team project.

What Does Ali Do?

Overview

This document lays out a plan for my work over the next year. My primary focus is the success of the retail sales engineering group. I expect to spend about half my time on technical direction for that group, and about 30% contributing to the NewMerchandising project, with the remainder split between cross-organizational initiatives (API working group, architecture reviews) and community work (interviewing, mentoring senior engineers). As part of the incident commander rotation, I expect to be on call 1 week out of every 10.

Goals

1. Make retail sales successful by guiding technical direction, contributing to org goal setting, and anticipating risks.
2. Act as a consultant/force multiplier for the success of NewMerchandising. Identify risks or gaps in engineering practices that threaten the project's goals.
3. Lead architecture reviews for teams in retail sales engineering.
4. Improve cross-engineering planning by participating in architecture reviews for other sales groups.
5. Act as extra leadership bandwidth when needed, such as during incidents or conflicts.

Sample activities

- Propose OKRs that address risks and opportunities for retail sales.
- Agree on goals and deliverables for NewMerchandising, and make sure teams are aligned.
- Consult on architecture for teams across the org. Recommend architectural approaches and contribute sections to RFCs, but unlikely to be primary author on any.

- Mentor/coach senior engineers.
- Interview senior and staff engineer candidates.

What does success look like?

- Retail sales is building systems that will scale for the next five years.
- The NewMerchandising project is making consistent progress with shared understanding of goals across all four teams.

Don't obsess about getting this perfect: get it *right enough*. Describing your goals doesn't mean you're forbidden from doing something else. But it's a nice reminder of what you intended to do, and it helps you keep an eye on whether you're actually doing the thing you claimed was your job.

You might decide that your focus needs to change earlier than you expected. The state of the world can change or your priorities might shift. If so, write a new role description with the new information. Being clear about your expectations of yourself makes sure everyone's on the same page.

IS THAT YOUR JOB?

Your job is to make your organization successful. You might be a technology expert or a coder or affiliated with a specific team, but ultimately your job is to help your organization achieve its goals. Senior people do a lot of things that are not in their core job description. They can end up doing things that make no sense in *anyone's* job description! But if that's what the project needs to be successful, consider doing it.

Some of my coworkers at Squarespace tell the story of the day in 2012 when their data center had a power outage and they carried fuel up 17 flights of stairs (*https://oreil.ly/6TZ2Q*) to keep it online. "Hauling barrels of diesel" does not show up in most tech job descriptions, but that's what was needed to keep the site online (and it worked!). When the machine room flooded at the ISP I worked at years ago, the job became about making a bucket chain of trash cans to keep the water level low. And when a Google project in 2005 was running late and we didn't have enough hardware folks available, my job for a couple of days was racking servers in a data center in San Jose. You do what you need to do to make the project happen.

Usually this "not my job" work is less dramatic, of course. It can mean having a dozen conversations to unblock a project your team depends on, or noticing that your new engineer is lost and checking in with them. To reiterate: your job is ultimately whatever your organization or company needs it to be. In the next chapter, I'll talk about how to understand what those needs are.

To Recap

- Staff engineering roles are ambiguous by definition. It's up to you to discover and decide what your role is and what it means for you.

- You're probably not a manager, but you're in a leadership role.

- You're also in a role that requires technical judgment and solid technical experience.

- Be clear about your scope: your area of responsibility and influence.

- Your time is finite. Be deliberate about choosing a primary focus that's important and that isn't wasting your skills.

- Align with your management chain. Discuss what you think your job is, see what your manager thinks it is, understand what's valued and what's actually useful, and set expectations explicitly. Not all companies need all shapes of staff engineers.

- Your job will be a weird shape sometimes, and that's OK.

2

Three Maps

As a staff engineer, you need a broad view. Every time you react to an incident, run a meeting, or give advice to a mentee, you'll need context about the people you're working with and what the stakes are. When you propose a strategy or move a project along, you'll want to understand how your organization works and the difficulties you might run into along the way. And you won't make good choices about what to work on unless you can step outside your day to day and see where you're all supposed to be going.

In Chapter 1, we zoomed out and took a big-picture view of what staff engineers are and why organizations need them. We defined some axioms that are helpful in understanding staff roles, and then I invited you to do a fact-finding mission to unpack some aspects of your own role: your *reporting chain*, your *scope*, your *work preference*, and your current *primary focus*. If you didn't already have a big picture of what your job is, I hope you now do. But if you've ever been hiking or navigated through a new city, you'll have seen that knowing where *you* stand is just the beginning. Getting oriented means knowing about your surroundings, too.

Uh, Did Anyone Bring a Map?

In this chapter, we're going to describe the big picture of your work and your organization by drawing some maps. Maps take different forms depending on their purpose: you wouldn't try to include elevation, voting districts, and subway navigation on a single map, for example. So rather than overlaying all of the information we have into one dense, unreadable picture, we're going to set out to build three different maps. They won't be perfect models, but they're useful tools for thinking about work and asking yourself questions about where you are, how your organization works, and what you're all trying to do.

You can approach this as a mental exercise—just a metaphor for thinking about your engineering organization—or you can actually set out to draw these maps. It can be enlightening (and fun) to compare notes with a colleague and see which landforms and points of interest you disagree on.

Here are the three maps we're going to end up with:

A LOCATOR MAP: YOU ARE HERE

We're going to start with your place in the wider organization and company. Last chapter we talked about your *scope*, but to truly understand that scope, you need to see what's outside it. What's along the borders? When you zoom way out, how big is your part of the world compared to everywhere else? Think of it like one of those maps that a news station throws up behind the presenter to remind you where a particular place is, and put it in context.

You need the locator map because it's tricky to be objective about any work while you're deep inside it. Unless you can maintain perspective, the concerns and decisions of your local group will feel more important to you than they would if you looked at them on a bigger scale. So we'll try out some techniques for getting that perspective. You'll be honest with yourself about which of the projects you care about would actually show up on a big map of the company, and which ones you wouldn't see unless you zoomed all the way in.

A TOPOGRAPHICAL MAP: LEARNING THE TERRAIN

The second map is all about navigating the terrain. If you're setting off across the landscape, you'll go further and faster if you have a robust knowledge of what's ahead. In this section, we'll look at some of the hazards on the map: the canyons and ridges along the fault lines of your organization, the weird political boundaries in places nobody would predict, and the difficult people everyone's been going out of their way to avoid. If there's quicksand ahead, or krakens to be wary of, or an impassable desert full of the sun-bleached skeletons of previous travelers, you'll want to mark those pretty clearly before you set out on your journey.

Despite the dangers and difficulties, you might find that there are navigable paths already in place. Discovering these paths will include understanding your organization's "personality" and how your leaders prefer to work, clarifying how decisions are made, and uncovering both the official and the "shadow" organization charts.

A TREASURE MAP: X MARKS THE SPOT

The third map has a destination and some points on a trail to get there. It shows where you're going and lays out some of the stops on the journey. The voyage might be perilous, but if you have a map, you'll be able to see whether you're getting any closer to that huge red X.

Uncovering this map means taking a long view and evaluating the purpose of your work. Is each project a goal in itself, or is it just a milestone along the path to the actual goal? Sometimes you'll discover that there isn't a destination at all or that there are several incompatible ones. When nobody has declared what the treasure is, or everyone disagrees on how to get to it, a staff engineer can have a huge impact by creating a vision or strategy, making decisions, or otherwise drawing a brand-new treasure map for the organization. But I'm getting ahead of myself. For now we're looking at uncovering the *existing* big picture. Creating a *new* one will happen in Chapter 3.

CLEARING THE FOG OF WAR

These three maps already exist in your organization; they're just obscured. When you join a new company, most of the big picture is completely unknown to you. A big part of starting a new job is building context, learning how your new organization works, and uncovering everyone's goals. Think of it like the fog of war (*https://oreil.ly/P6S9K*) in a video game, where you can't see what awaits you in the parts of the map you haven't explored yet. As you scout around, you clear the fog and get a better picture of the terrain, learning what's surrounding you and whether there are wolves coming to bother your villagers. You can set out to uncover the obscured parts in all three of the maps and find ways to make that information easy for other people to understand. For instance:

- Your locator map can help you make sure the teams you work with really understand their purpose in the organization, who their customers are, and how their work affects other people.

- Your topographical map can help highlight the friction and gaps between teams and open up the paths of communication.

- Your treasure map can help you make sure everyone knows exactly what they're trying to achieve and why.

You'll be able to clear some parts of the map through everyday learning, but you'll need to deliberately set out to clear other parts. A core theme of this chapter is how important it is to know things: to have continual context and a sense of what's going on. Knowing things takes both skill and opportunity, and you might need to work at it for a while before you start seeing what you're not seeing.

Let's start with the skill. I spent a few months in the Irish countryside during the pandemic and went for a lot of nature walks with my friends who live there. At first, I thought I was seeing everything there was to see: a bunch of foxgloves or an oak tree, things that were striking and beautiful. But my friends were seeing more than I was. They'd pause at a patch of mud that I wouldn't have looked at twice and point out the footprint of a pine marten. They'd pick out leaves that I would have dismissed as just grass, and note that they're delicious and peppery, a treasure for foragers. Even the kids would see little flowers or dive on a patch of wild strawberries that I'd have walked right past. Why could they see all of these things when I couldn't? Because they had learned to pay attention and they knew what they were looking for.

Paying attention means being alert to facts that affect your projects or organization. And that means continually sifting information out of the noise around you. If you can train your brain to say "That's interesting!" and remember facts that you might need later on, you'll start to add detail to your maps and build skills in synthesizing new information.

What sorts of facts are useful? Anything that can help you or others have context for your work, navigate your organization, or progress toward your goals. Here are some examples:

- A company all-hands presentation about an upcoming marketing push might be a hint that huge traffic spikes you're not ready for are coming your way.

- Your director asks you to take on a project you don't have time to do, but you know which senior engineers in your organization are ready for opportunities to stretch their skills.

- A shift in corporate priorities could mean a platform you'd considered but backburnered has become an amazing investment.

- Your database just disappeared, and you remember getting an email about network maintenance.

Over time, you'll get used to how news travels in your org and what you should pay attention to. You'll know which emails you need to read and which meetings you need to go to. If your brain's not naturally "sticky" for retaining information like this, I recommend challenging yourself to note down facts that might be useful later, just to get yourself into the habit of paying attention. Think of gathering context as a skill to build as part of your job.

But noticing only takes you so far. Paying attention doesn't help if you don't have access to the decisions and discussions that affect your work. While you might be privy to a daily flow of meetings, emails, and regrettable @here messages on Slack, there's a lot of other information you won't be able to ask for unless you know it exists. How do you get into the "room where it happens"? I'll share some strategies in this chapter.

The Locator Map: Getting Perspective

As you grow in seniority, making a real impact will mean being able to put your work in a bigger context, and recognizing that your point of view is heavily influenced by where you're standing. (Figure 2-1 gives maybe a little too much perspective.)

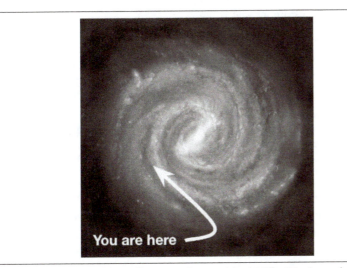

Figure 2-1. A locator map of the Milky Way galaxy (original Milky Way image by Jean Beaufort, CC0).

Of course, everyone else you work with will have their own point of view too: their "You are here" marker will be somewhere else on the map. If you want to make good decisions, you'll need to be able to see from some of those other points of view.

The more time you spend absorbed in any domain and learning the nuances of the work at your scope, the richer and more complex it will become for you. As you understand the people, the problems, and the goals, you'll become more focused on them. That focus brings depth and understanding, but it comes with some risks, especially for a staff engineer.

Let's look at four of those risks now.

Prioritizing badly

When everyone around you cares about the same set of things, it's easy to magnify the importance of those things. The problems that exist outside your group can start to appear simple or unimportant by comparison. That's why you see teams making those local maximum decisions I talked about in Chapter 1: the local maximum starts to feel *really* important. The more time you spend staring at your own group's problems, the more they seem special and unique and worthy of special, unique solutions. And sometimes they are! But it's unusual to find a problem that is genuinely brand new. If you check for prior art and preexisting solutions, you'll spend less time reinventing wheels.

Losing empathy

It's easy to overfocus and forget that the rest of the world exists, or start thinking of other technology areas as trivial compared to your rich, nuanced domain. It's like you start looking at the world through a fish-eye lens that makes the thing right in front of you huge and squeezes everything else into the periphery. You can lose empathy for the work other teams are doing: "That problem they're solving is easy. I could solve it in a weekend."

The words you use, the things you choose to explain versus those you leave implicit, and the motivations you ascribe to other people will all be influenced by your perspective. That's why it can be so difficult for engineers to communicate with nonengineers. Figure 2-2 tells a familiar story about how easy it is to misunderstand what other people know about your domain.

Loss of empathy shows up in incidents, too, where teams can get absorbed in the interesting technical details of the problem and forget there are users waiting for the system to be back online.

Figure 2-2. It's easy to lose perspective about what other people know (source: https://xkcd.com/ 2501 by Randall Munroe).

Tuning out the background noise

If one failure mode is your team's concerns seeming more important than everyone else's, another is the exact opposite: you stop noticing problems at all! If you've been working around the same mucky configuration file or broken deploy process for months, you might get so used to it that you stop thinking of it as something you need to fix. Similarly, you might not notice that something that started out as just slightly annoying has slowly become worse. Maybe a problem is close to becoming a crisis, but you don't even notice it anymore, so you can't be objective about how quickly you need to react.[1]

---

1 A popular metaphor, the boiling frog, says that if you drop a frog into a pot of boiling water, it will jump out, but if you put a frog into cold water and very gradually increase the temperature, you can bring the water to a boil and kill the frog. It's often used as a cautionary tale to illustrate that gradual change can become normal and that we can slide into catastrophe without reacting to it. I was so relieved when I learned that real frogs don't behave like this: they just jump out! Let's leave the poor frogs alone, but the metaphor is useful.

Forgetting what the work is for

Being in your silo can mean that you lose your connection to what's going on elsewhere in the company. If your group originally took on some project to solve a larger goal, the project might still be ongoing even though the goal no longer matters or has already been solved in some other way. If you're working only on your own little part of a project, it's easy to stop thinking about what the project is *for*. You can slip into a world where everyone does their own little part and nobody feels like they're responsible for the end result. You can lose sight of the ethics of what you're doing, too, and find yourself working on something that you wouldn't really be OK with if you stepped back and thought about the whole picture.

SEEING BIGGER

Open up your company's org chart and look at where your group and others you care about connect to the rest of the organization. When you extend the amount of the map you can see, your own group might seem a lot smaller, and your "You are here" pin might feel far from where the action is. But without perspective, you can't do impactful work. In this section we'll look at some other techniques for seeing the bigger picture

Taking an outsider view

When I was the newest person on an infrastructure team years ago, my colleague Mark commented after a few weeks, "There's this facial expression you do when I describe our systems…" Certainly I'd thought a few of the aged systems needed to be replaced, but I hadn't realized I was wearing my opinions so clearly (and rudely!) on my face. Two years later, the team's hard work meant that the architecture had vastly improved. We were proud of the work. I thought it was pretty good! Until a new person joined and…wore their opinions pretty clearly on their face. By then, I had become a team insider. I needed a newer "new person" to help me see the problems again.

When the new person on your team looks at an architectural tangle or a pile of technical debt, they have no historical context. As my colleague, Dan Na, says (*https://oreil.ly/GD8Gz*), a new person can always see the problems. They haven't been around for the gradual change and the boiling frogs: they're just seeing the raw situation as it is. Without preconceptions, they're free to look around and ask, "What's really happening here? Is any of this working?"

Warning

Being new isn't a license to be a jerk. It's easy with hindsight to say, "This is terrible! Why didn't they just..." But have humility and assume there are good reasons for everything being the way it is. Amazon's principal engineer group acknowledges this in one of its community tenets: "Respect what came before" (*https://oreil.ly/2R4ET*).

Being new is the best opportunity you'll have to get a complete outsider view, but as a staff engineer, you should try to have this perspective all the time. You need to be able to look at your own group as if you weren't part of it and to be honest about what you see. Do your technical decisions only make sense to people who have forgotten that there's a world outside your team? If you all stopped doing the work you're doing, how long would it be before other people would notice or care? Have you gotten absorbed in the technology and forgotten what your original goal was? *How is everything?* The next four sections offer techniques for viewing things like an outsider.

Escaping the echo chamber

When you find yourself in an echo chamber where everyone you meet holds the same set of opinions, it can be a shock when you connect with peers in other groups and discover that some of their views are just...different.

After spending more than a decade at the bottom of the stack in infrastructure roles, it was a shock to my system when I first worked with product engineering teams. They moved *fast*, took risks, and thought creating features that customers loved was *at least as important* as those features having rock-solid reliability. Our debates shook some of my firmly held beliefs and made them more nuanced.

Seeking out peers in other groups is an important part of your job. Build friendly relationships with other staff engineers. Get to a point where you can speak the truth to one another, and it won't be contentious, because you've built up so much goodwill. This includes understanding any negative opinions that other teams hold about your group—if you start seeing what's valid about their comments, you'll do better work. Think of the other staff engineers as *your team*, just as much as any team you're part of.

The same principle applies across organizations. In Figure 2-3 and Figure 2-4, I depict each staff engineer as scoped to a single group, and each principal engineer as scoped to an organization. While the actual structure will vary, the point is to be part of something that's bigger than your own team or group, so you can have a more objective view of what everyone is doing.

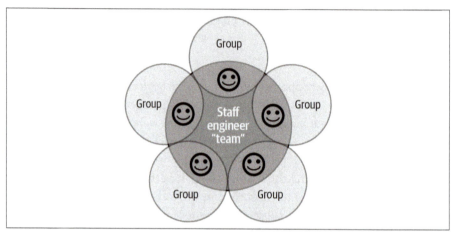

Figure 2-3. An example software engineering organization. Each group here contains multiple teams. In this company, each staff engineer's scope is a single group, and they consider themselves to be part of their own group, but also part of a bigger virtual "team" of staff engineers.

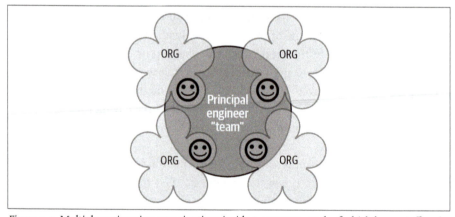

Figure 2-4. Multiple engineering organizations inside a company, each of which has a staff engineer. Every principal engineer is in their own org, but is also part of the virtual team of principal engineers.

Go beyond engineering: build relationships with product folks, customer support, administrative staff, and more. If your work affects them or their work affects you, go be friendly and understand their point of view. It will give you a whole new way of thinking about what's important to your department or your business.

What's actually important?

Befriending nonengineers is good for your perspective in another way: As an engineer, it's easy to get absorbed in technology. But technology is a means to some end. Ultimately you're here to help your employer achieve its goals. You should know what those goals are. You should know what's *important*.

A startup will have a different definition of what matters than a behemoth tech giant or a local nonprofit. A mature product will have different needs than an early one. Some goals, and thus some projects, matter more than others. Figure 2-5 shows how a project that feels like the center of your universe can be much less significant when looking at a bigger picture. The ordering will change over time, so understand what matters *right now*. If your customers are leaving in droves because your product is missing core features that your competitors have, it's probably not the time to push for a focus on technical debt. If everything is smooth sailing and you're anticipating growth, this might be a great time to make sure your foundations are solid.

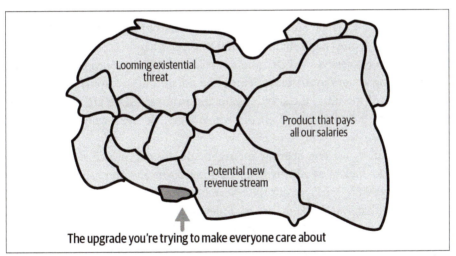

Figure 2-5. *Putting your project in perspective. That upgrade might be the most important work your organization has, but people looking at a bigger picture won't see it as important.*

A company's goals extend beyond its stated objectives and metrics; they include "continuing to exist," "having enough money to pay everyone," and "having a good reputation." My colleague, Trish Craine, head of operations for engineering at Squarespace, calls these "the objectives that are always true." These are the needs of your company that are so obvious, they're only really stated if they're in danger. The product or service that your organization provides should *work*. Its customers should want to use it. Deploying it shouldn't be painfully slow. Know your implicit goals as well as the explicit ones.

> **Tip**
>
> As time passes, your company's priorities will change and parts of your map will fog up again. To stay up to date with what's important, pay attention to all-hands meetings for your group and others, ask for skip-level one-on-ones with your manager's manager, and find face time with customers or teams that depend on you. If you don't have business context about why (or whether) your work matters, ask for it.

Notice when the goals change, too, because that might mean your scope or focus should change. It's OK if you're not working on the *most* important thing, but what you're doing should not be a waste of your time. If you can't explain to yourself why what you're doing needs a staff engineer, you might be doing the wrong thing.

What do your customers care about?

Charity Majors, CTO of Honeycomb, often hands out stickers (*https://oreil.ly/2pxrj*) that say: "Nines don't matter when users aren't happy." "Nines" here refers to service level objectives (SLOs) (*https://oreil.ly/9LeCI*), a common mechanism for measuring system availability. "Three and a half nines of availability" means that 99.95% of the time, the service is up and running. SLOs are useful, but as Majors points out, they don't tell the whole story. Because who defines what "available" means?

Mohit Suley, an engineering manager and former principal engineer at Microsoft, has spoken about his team tracking down and contacting unreliable ISPs (*https://oreil.ly/Fsj4k*) where their search engine, Bing, wasn't reachable. It wasn't Bing that was broken, but as Suley says, "A user doesn't distinguish between DNS services, ISP, your CDN, or your endpoint, whatever that might be. At the end of the day, there are a bunch of websites that work, and a bunch that don't." You need to measure success from your users' point of view. (If your customers are other teams inside your company, this still applies!) If you don't understand your customer, you don't have real perspective on what's important.

Have your problems been solved before?

Amazon's tenet of "respect what came before" (*https://oreil.ly/2R4ET*) includes a reminder that "many problems are not essentially new." You'll come up with better solutions if you study what other people have already done before creating some new thing. Remember that your goal is to solve the problem, not necessarily to *write code* to solve it. Take the time to understand what already exists—inside and outside your organization—before building something new.[2]

Industry Perspective

Understand how other people in the industry have solved the problems you're working on. Your preferred publications and resources will depend on your interests, but here's some I find valuable for architecture, technical leadership and software reliability. I love the LeadDev (*https://oreil.ly/P3SYi*) and SREcon (*https://oreil.ly/S6OYy*) conferences and try to make it to as many as I can. LeadDev has a new (at the time of writing) conference track called StaffPlus (*https://leaddev.com/staffplus-new-york*). I'm hosting some of their events, so I can't be entirely objective, but I think it's excellent!

For online conversations, I like Rands Leadership Slack (*https://oreil.ly/ZheFA*): the #architecture and #staff-principal-engineering channels are gold. The LeadDev Slack's #staffplus channel is very active during events, too.

I subscribe to the monthly InfoQ Software Architects' Newsletter (*https://oreil.ly/ReBFX*), as well as the VOID report (*https://oreil.ly/wx82Q*), and SRE Weekly (*https://sreweekly.com*). I read the Raw Signal newsletter (*https://oreil.ly/CwcQp*) for a weekly dose of a manager point of view. I also eagerly await the quarterly Thoughtworks Radar (*https://oreil.ly/iuOSy*).

Whatever domain you're in will have its own publications. Use them to maintain perspective and spot new ideas that you can explore when you need them. They'll help you keep learning, too.

---

2 That's also why design documents should have an "alternatives considered" section; we'll talk more about design docs in Chapter 5.

The Topographical Map: Navigating the Terrain

A locator map gives perspective, but you can't navigate by it. You need another map: one that shows the terrain.

Geologists study *plate tectonics*, the way the huge pieces of the earth's lithosphere (see Figure 2-6) move against each other over time, forming mountains and trenches and creating earthquakes and volcanic activity. Team tectonics have similar properties. As domains of responsibility smash against each other, they form an organizational terrain, complete with overlaps and conflict, ridges and chasms.

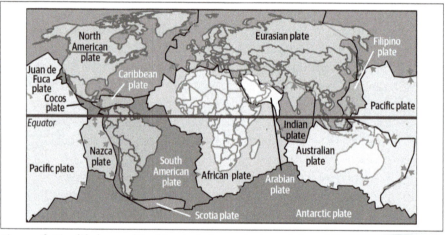

Figure 2-6. Simplified map of Earth's principal tectonic plates. (adapted from: Scott Nash, public domain, https://oreil.ly/UdeNz).

Reorganizations can disrupt communication between groups that need to work closely together. Teams that are under a heavy load can entrench and put up barriers. A new senior leader can cause an earthquake that reshapes the landscape overnight. Navigating an organization (see Figure 2-7) requires a *topographical* map.

Figure 2-7. A staff engineer navigating tricky terrain.

ROUGH TERRAIN

Let's explore some of the difficulties you'll face if you set out on a mission without a detailed map of the terrain.

Your good ideas don't get traction

Being *right* about a need for change is less than half the battle. You'll have to convince other people that you're right and, even more difficult, convince them to *care* that you're right. That means knowing how to build momentum within your organization: figuring out who can sponsor your idea or help it spread, and how you can get it over the finish line and make it "real."

You don't find out about the difficult parts until you get there

Many obvious-seeming journeys have some crucial point that nobody has figured out how to get past. You may be attempting to scale a cliff that's defeated many other people before. Staff engineers can often navigate past obstacles that less experienced engineers can't, and it's possible that you'll be able to succeed where others have failed. But if you know where people got stymied in the past, you can take a different path or solve the hardest part of the problem first, so other people will be convinced the project is worth their effort.

Everything takes longer

Unless you know how your organization works, decisions that should be straightforward will take months or quarters. The mechanics of your organization's planning cycles will affect you too. There are times of year when it'll be easier to make the case for staffing a new project or to rally everyone behind some goal. If you announce an initiative immediately after the quarterly engineering OKRs have been set, you'll have a hard fight and you may have to wait a quarter before you'll see any progress toward your goal.

UNDERSTANDING YOUR ORGANIZATION

Engineers sometimes dismiss organizational skills as "politics," but these skills are part of good engineering: considering the humans who are part of the system, being clear about the problem you're solving, understanding long-term consequences, and making trade-offs about priorities. If you don't know how to navigate your organization, every change will be much more difficult.

In this section, I'll describe some ways you can clear the fog and understand your company's terrain. That starts with evaluating some aspects of your culture, including what gets written down, how much trust there is, whether people are eager or hesitant to change, and where new initiatives come from. This knowledge will set your expectations about an average journey: will it be easy to make progress? After that, we'll look at some of the obstacles and shortcuts that will show up on your topographical map.

What's the culture?

Whenever I interview a job candidate, their first question is often, "What's the culture like?" I used to struggle to answer; where do you even start? Tomes have been written on organizational culture. Now, though, I think most of the time people are really asking these questions:

- How much autonomy will I have?
- Will I feel included?
- Will it be safe to make mistakes?
- Will I be part of the decisions that affect me?
- How difficult will it be to make progress on my projects?
- Are people...you know...*nice?*

Your company culture is not the only factor determining the answers to these questions: individuals and leadership are part of it, too. But organizations do have their own distinct "personalities," so let's talk culture.

If your organization has published a statement of values or principles, that can help you see what the leaders care most about. But these values are aspirational: the real values of the company are reflected in what actually happens every day.

To understand more about the engineering culture at your company, here are a few questions you could ask yourself or discuss with a colleague. For most of them there's no right or wrong answer. It will be difficult to maneuver if your company is all the way over on one side or another, but there's a lot of space for success in between.

Secret or open? How much does everyone know? In secret organizations, information is currency and nobody gives it away easily. Everyone's calendars are private. Slack channels are invite-only. Often you can get access to something if you ask for it, but you have to know it exists! When all information is need-to-know, it's harder to come up with creative solutions or really understand why something's not working.

In open organizations, you'll have access to *everything* (even messy first drafts!). You might get decision fatigue from choosing which information to consume. You might not know which documents are official and need action, and which are just early ideas. And open information can lead to more drama: it's harder for bad ideas to be quietly shut down.

Knowing the cultural expectations around sharing is crucial. In a culture that keeps knowledge locked down, you'll lose your boss's trust if you reshare something they told you in confidence. In a more open company, you'll be considered political or untrustworthy if you withhold information or don't make sure everyone knows what's going on.

Oral or written? What gets shared by word of mouth and what gets written down? How much writing and review is involved in decisions? In some companies, it's typical to make a big decision during a hallway conversation or to find out your colleague has built a huge new feature after it launches (or when you get paged for it). In other workplaces, every software change comes with a formal specification, requirements, sign-offs, and an approvals checklist, and you can expect a one-line change to take a quarter.

Thankfully, most workplaces are somewhere in between. If yours prefers quick conversations, you may get pushback if you take the time to write a decision down—and a design document longer than a page just won't get read. Bigger and more mature companies tend to be more deliberate about changes. If you're at one of those and you *don't* create a change management ticket or a design document, you'll seem sloppy and irresponsible. One team I worked in had a cowboy hat that would end up on the desk of whichever team member had last done something a little too "Wild West." It was affectionate, but it was a good reminder too.

Top-down or bottom-up? Where do initiatives come from? A completely bottom-up culture is one where employees and teams feel empowered to make their own decisions and champion the initiatives they think are important. However, when those initiatives need broader support, they slow down. If teams disagree about direction or priority, the lack of a central "decider" can lead to deadlock.

On the other hand, people in a fully top-down company will find it much easier to choose initiatives and take decisive action. Those decisions won't be the best ones, though, because they're missing local context. The engineers likely feel controlled and won't be empowered to react to changes as they arise.

Staff+ engineers should be fairly autonomous and self-directed, but make sure your organization agrees: if your manager expects to approve where you spend your time, it can cause conflict if you don't check in. If you're used to seeking permission or having work handed down, and you move to a bottom-up company, you'll be seen as having low initiative and have trouble getting anything done.

If you know how your organization tends to work, you'll also know whether to take your ideas to fellow grassroots practitioners and get their support first, or whether to start by trying to convince your local director.

Fast change or deliberate change? Younger companies tend to make rapid decisions and pivot abruptly to try a new opportunity. As companies get larger and older, they take longer to change course. "Fast" organizations may be repelled by the idea of taking on a long-term project like a two-year migration. Slow ones will miss low-hanging opportunities to improve.

Depending on where you are, you'll need to frame your initiatives differently. If you're somewhere that moves like lightning, you'll want an incremental path that shows value immediately. In a more deliberate environment, you'll need to show that you've thought through the whole plan. This is tightly connected to oral and written culture, too.

Back channels or front doors? How do people in different groups talk with each other? There may be formal paths for information and requests, but your social culture adds informal channels too. If people are friendly across teams, they'll send a DM when they have a question and share ideas over coffee.[3]

If an engineer in one group can just go chat with a counterpart in another, it's going to be easier to make decisions that cross both teams. In some places, the only real way to get work done is to have an "in" via a back channel with someone on the team. If it's more typical to file a ticket and wait, or to send a collaboration idea up your management chain until you and the other team have a manager in common, everything will take longer—but it will also be more predictable and fair.

Understand what's considered typical in your organization. If everyone is strict about only using formal channels, it will be considered rude to ask questions out of turn, and people will judge you poorly for skipping the queue. If back channels are the typical way to get things done, you'll be waiting for a response for a month when you could have just had a chat with that person who's been admiring your cat pictures on the company pets mailing list.

Allocated or available? How much time does everyone have? If teams are understaffed and overworked, you'll have trouble finding a foothold for any new idea that isn't on an existing product road map—the fastest and easiest response is just to say no without really looking at the request. You'll have the most impact with any initiative that can free up time without major investment. Your most likely successes will come when you can work alone or with a little help, and don't need to get a bunch of busy people to commit to anything new.

Teams that aren't busy may seem easier to work with, but they have a different problem: underallocated engineers rarely stay that way for long. If there are plenty of free cycles available, chances are that a Cambrian explosion of competing novel grassroots initiatives is taking hold, each with a small number of devotees. You'll have more impact if you choose a nascent project, help it over the finish line, and convince others to rally around it too.

Liquid or crystallized? Where do power, status, and reputation come from? How do you gain trust? Some organizations, particularly in academia and in big and old companies, have a clear hierarchy: the same group of people, in the same

---

3 Shared-interest Slack channels, social clubs, and employee resource groups (ERGs) can be fantastic ways to get to know people and make connections across the organization. Shoutout to my friends on the #crosswords channel at work who share their New York Times Crossword times every day and the #women-in-engineering channel participants who celebrate every member's successes.

configuration, climb the ranks together and have a fairly fixed structure for communicating, making decisions, and allocating the "good projects." Each person is like a node in a crystal lattice: so long as the people around you are moving up, you'll move up too. Senior people in groups like this will often say that they never looked for promotion: they stayed where they were, got a project, got support from the group, and got the promotion when it was their turn.

This sort of hierarchy is anathema to young, small, scrappy companies, which claim to be something like a meritocracy. Let's be realistic about that (*https://en.wikipedia.org/wiki/Myth_of_meritocracy*): success still depends on having access to opportunities and sponsorship, so it's hugely affected by stereotype bias, in-group favoritism, and other cognitive biases. (See the website Is Tech a Meritocracy? (*https://istechameritocracy.com*) for more.) "Liquid" companies offer more room to change your place in the structure, but you'll likely have to hustle a bit to get promoted. You might move from group to group to find high-impact work so you can advance at your own pace. If you sit around waiting for someone to assign you a project, you'll be waiting a long time.

In teams with a solid crystalline lattice, it's vital that you understand your place in the hierarchy and know when your time will come to have a project that will take you to the next level. If you suggest taking on something that's been earmarked as a promotion project for someone else, you'll ruffle feathers—one friend who tried this said their boss looked at them "like I'd suggested stealing the silverware."

Try the slider diagram in Figure 2-8 to think about how these seven attributes influence how your organization works. If you're trying to cause a culture change, it's often possible—with determined effort—to nudge the sliders in one direction or the other over time. At least know where they are, and you'll avoid some of the pitfalls of working against the prevailing culture.

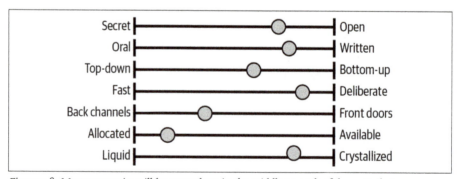

Figure 2-8. Most companies will be somewhere in the middle on each of these attributes.

Power, rules, or mission?

Here's another lens to view your culture through: what do your leaders think is important? In his 2005 paper, "A Typology of Organisational Cultures" (*https:// oreil.ly/rtHz5*), sociologist Ron Westrum wrote:

> *Through their symbolic actions, as well as rewards and punishments, leaders communicate what they feel is important. These preferences then become the preoccupation of the organization's workforce, because rewards, punishments, and resources follow the leader's preferences. Those who align with the preferences will be rewarded, and those who do not will be set aside. Most longtime organization members instinctively know how to read the signs of the times and those who do not soon get expensive lessons.*

Westrum classified organizations and their influence on information flows using three categories (see Table 2-1):

Pathological
A low-cooperation culture where power and status are the goal and people hoard information; in Westrum's words, "a preoccupation with personal power, needs, and glory"

Bureaucratic
A rule-oriented culture where information moves through standard channels and change is difficult; the preoccupation here, Westrum writes, is with "rules, positions, and departmental turf"

Generative
A mission-oriented, high-trust, high-cooperation culture where information flows freely; Westrum calls this "concentration on the mission itself"

Table 2-1. The Westrum organizational typology model: How organizations process information (Ron Westrum, "A typology of organisational cultures" (https://oreil.ly/tUnoa)) BMJ Quality & Safety 13, no. 2 (2004), doi:10.1136/qshc.2003.009522.

Pathological	Bureaucratic	Generative
Power-oriented	Rule-oriented	Performance-oriented
Low cooperation	Modest cooperation	High cooperation
Messengers shot	Messengers neglected	Messengers trained
Responsibilities shirked	Narrow responsibilities	Risks shared

Pathological	Bureaucratic	Generative
Bridging discouraged	Bridging tolerated	Bridging encouraged
Failure → scapegoating	Failure → justice	Failure →inquiry
Novelty crushed	Novelty → problems	Novelty implemented

The DevOps Research and Assessment (DORA) group (now part of Google Cloud) has shown (*https://oreil.ly/Epx4s*) that high-trust cultures that emphasize information flow have better software delivery performance. It's not surprising that an increasing number of software companies aim to have a generative culture. That means encouraging cooperative cross-functional teams, learning from blameless postmortems, encouraging experimentation, taking calculated risks, and breaking down silos. If your organization works like this, you're going to have an easier time sharing information and making progress.

If you know whether your workplace is oriented around power, rules, or mission, you'll find it easier to get things done. A feeling for how much people will share information, cooperate, take the time to help, and get behind new ideas will keep you safer and less frustrated as you cross the terrain. If you know you're in a bureaucracy (*https://oreil.ly/4zT4y*), you'll have more success if you plan ahead, stay within the rules, and respect the chain of command. If you're in a pathological organization, you'll take fewer risks—and cover your ass when you do. Pushing a cart across cobblestones is more difficult than doing so across smooth paving. If you know the road will be rocky, you'll budget more time, and you're less likely to get mad at aspects of the situation that you can't control.

Noticing the points of interest

This brings us back to the topographical map. Understanding your organization's culture gives you a rough idea of how easy or difficult an average journey will be. But to navigate, you'll also want to understand the barriers, the difficult parts of the journey, and the shortcuts. Here are a few points of interest in the terrain of organizations I have known.

Chasms The plate tectonics of a company can form chasms between teams and organizations. For instance, Figure 2-9 depicts the canyon that can form between product-focused software engineering teams and the infrastructure, platforms, or security teams providing services for them. Group cultures, norms, goals, and expectations evolve differently, causing gaps that make it difficult to communicate, make decisions, and resolve disputes.

Figure 2-9. The chasm between an infrastructure and a product engineering organization.

Smaller chasms can form even within an organization. The edges of each team's defined responsibilities rarely line up perfectly, and project work and information can get lost in the gaps between teams.

Fortresses Fortresses are teams or individuals who seem determined to stop anyone from getting projects done. Maybe you need their approval but can't get time with them. Or maybe they're gatekeepers and seem to decide your idea is bad before they even know what you're asking. Although some fortresses are petty tyrants, the majority are well-intentioned. They gatekeep because they *care*. They're trying to keep the quality of the code or architecture high and keep everyone safe.

To pass through the fortress gates, you might need to bring a token of sponsorship from someone the gatekeeper respects, or know the password to lower the drawbridge. (Common passwords include proving that you've mitigated all of the risks of your proposed change, completing lengthy checklists or capacity estimates, or replying to huge numbers of document comments with acceptable answers.) Another option is a protracted, bloody battle where you argue every point and pull other people into the fight—winning one of those can be such a Pyrrhic victory that you almost regret trying. Or you can give up and go the long way around the fortress, complicating your journey and losing access to any wisdom the gatekeeper would have shared.

Disputed territory It's very hard to draw team boundaries in a way that lets each team work autonomously. No matter how opinionated your APIs, contracts, or team charters, there will inevitably be some pieces of work that multiple teams think they own, and navigating those disputes can feel risky.

I worked on a project once that needed a critical system to be migrated from one platform to another. Migrating this particular system accounted for less than

5% of my project, so I didn't want to spend too much time on it—but when I looked for someone to take responsibility for it, I hit a wall. Ownership of the system was smeared across three teams, each responsible for a different aspect of its behavior. Nobody could tell me whether migrating it to our new platform would be safe. Each group said, "Yes, as far as I know, but you should also ask..." and pointed to the next team. Without an owner who could speak for the whole system, I went around in circles trying to build enough context to convince myself that the migration would work. (It didn't. Aligning the three teams around the rollback wasn't pretty either.)

When two or more teams need to work closely together, their projects can fall into chaos if they don't have the same clear view of where they're trying to get to. The lack of alignment can lead to power struggles and wasted effort as both sides try to "win" the technical direction. Overlaps in team responsibilities make this worse, complicating decision making and wasting everyone's time.

Uncrossable deserts   As you try to achieve your goals, you'll sometimes run into a battle that other people consider unwinnable. This may be a project that's just too big or a politically messy situation that always ends with a veto from some senior person. Whatever it is, people have tried it before, and any suggestion of tackling it again will be met with discouragement and ennui.

That's not to say you shouldn't try! But you should have enough evidence to convince yourself and others that this time will be different. It's good to know going in if you're picking a fight that might be unwinnable.

Paved roads, shortcuts, and long ways around   Companies that have worked to make engineers efficient will often set up processes to ensure that the official ways to do something are also the easiest ways. An example might be following a self-service checklist to ensure a new component is safe to put into production. If you're lucky enough to have some of these easy, well-defined paths, know where they are and use them.

Unfortunately, not all roads are well paved. We've all tried to solve a problem the official way for a long time before someone told us the secret path to success: the undocumented search feature, the admin who can set up an account for you, or the one person in IT who responds to DMs. Sometimes the official way is the way that everyone learns *not* to use. Figure 2-10 shows a paved road that doesn't lead to most of the places people actually want to go; they take the legacy paths instead. If you don't know these goat tracks through your org, everything takes longer. But when you learn them, teams may ask you not to document them: people *should* use the new path, they insist, and it will be better *soon*.

Figure 2-10. The new paved road is beautiful, but most of the places people actually want to go are deep in the marsh.

WHAT POINTS OF INTEREST ARE ON YOUR MAP?

What else should be on your topographical map? Are there unexpected cliffs you can walk off? Are there behaviors or communication styles that would be perfectly fine in another company or team but that are considered rude in yours? Are there guardrails you expect to be in place that just aren't? Are there areas that are prone to eruptions, or leaders who cause earthquakes (or surprise reorgs) for people who thought they were on solid ground? How about local politics—which teams are led by monarchs, and which by councils? Which ones are anarchy? Who's at war with whom?

Try sketching your own map. Remember that cartography is inherently political (*https://oreil.ly/V5UbQ*): what you choose to include says something about *you* as well. Pay attention to what you put at the center of the map and where you're inclined to take sides.

Organizations end up with weird shapes due to reorgs, acquisitions, individual personalities, and, in some cases, people who just don't like each other. If you come up with barriers, conduits for information, or other landforms that I haven't thought of, I'd love to hear about them.[4]

How are decisions made?

It is fascinating to watch how information and opinions flow through a company and to see how unexpectedly they can become a plan of record. Suddenly everyone's using a new acronym or holding a particular opinion, and it can be hard to see where that came from. A project that held great hope and promise is now dismissed as likely to fail. Everyone's excited about microservices, or they've moved on from microservices and they're curious about serverless, or they think a modular monolith is just pragmatic common sense. One team has approval to hire more people this year and another doesn't. How did all of these decisions happen? Was there a memo?

Some decisions seem to emerge from conversations without anyone really declaring that they've decided. Others happen more formally, but in rooms you're not in. If you have a lot of ideas, it can be frustrating when you see other initiatives take root but not yours. Why aren't they listening to your proposals?[5] The truth is something that a lot of us struggle to make peace with: being technically correct about a direction is only the beginning. You need to convince other people too—and you need to convince the *right* people.

If you don't understand how decisions are made in your organization or company, you'll find yourself unable to anticipate or influence them. You might also find that you think you hold the same opinions as everyone else about what should happen next, and then find that suddenly everyone is advocating for a different path. If you consistently feel out of the loop, that's a sign that you don't understand how decisions are made and who influences them.

---

4 I can recommend discussing landforms with a fifth grader, if you have one in your life. They're well adapted for questions like "What would a fjord be if it was a metaphor for humans trying to work together?" (Two teams worked together to make a big project—a glacier—but they got angry with each other, the project melted, and all that's left is the water at the bottom. Now you know.)

5 The unspecified "they" often works as a keyword to alert you that you're operating without enough information. If you find yourself having a thought like this, double-check who you mean by "they." If it's "the whole organization," then that's part of your problem. Understand exactly who you need to convince. I'll talk more about the official deciders and the "shadow" org chart later in this chapter.

Where is "the room"?

Decisions that affect you and your scope are happening every day, and it's uncomfortable if you keep being shocked by them. You should at least have a feeling for where they're coming from, and you'll likely want to have some influence on them too. Let's start with the formal channels and the official meetings where big decisions get made.

Your access to decisions will be different depending on where in the organizational hierarchy you sit. Some of these decisions will inevitably be happening higher up in the company than you are. You can influence them by making sure relevant information reaches those rooms via your reporting chain or other channels. But decisions are also being made directly at your scope, and, as much as possible, you'll want to be involved. If you've watched the musical *Hamilton*, you'll remember Aaron Burr's craving to be "in the room where it happens." As Burr tells us, people who aren't in the room "don't get a say in what they trade away" (*https://oreil.ly/G1Csw*). While there are times when an outsider perspective can help, this isn't one of them. If you want to set technical direction or change your local culture, you need to be an insider in the group that's making the decisions.

Figure out where decisions are happening. Perhaps there's a weekly managers' meeting that's intended to make organizational decisions but that often weighs in on process or technical direction. A director might tend to make plans in their staff meeting with the people who report to them. A central architecture group might have a Slack channel where they come to consensus on the path forward. If you're not seeing how your organization works, ask someone you trust to walk you through where a particular decision came from. (Be clear that you're not fighting the decision, just trying to understand the inner workings of your organization.)

Beware: there might not be a "room" at all. At the most extreme ends, major technical pronouncements might get made in one-on-ones with the most senior leader, or they might be intended to be entirely bottom-up (and therefore often not made at all).[6] But if there is a "room where it happens" for the kind of decision you're interested in, find out what that is and who is in it.

---

6 "Coordination Headwind: How Organizations Are Like Slime Molds" (*https://oreil.ly/n2nNf*) is a fantastic presentation about the failure modes of bottom-up coordination.

Asking to join in

Once you discover a meeting where important decisions get made, it's natural to want to be part of it. But you'll need a compelling story for why that should happen. It seems obvious, but your reasons should be about impact to your *organization*, not to you *personally*. No matter how much your peer managers like you, framing your exclusion as being bad for your career advancement will be unlikely to change hearts and minds. Show how including you will make your organization better at achieving its goals. Show what you can bring that's not already there. Have a clear narrative about why you need access, practice your talking points, and go ask to join.

You will probably get some resistance. Adding someone to a group is rarely free for the people who are already there. Every extra person in any meeting slows it down, extends discussions, and reduces attendees' willingness to be vulnerable or brutally honest. If the group is used to working together, every new person resets the dynamic; to some extent, attendees have to learn to work together again.

If you do get an invitation, don't make anyone regret inviting you. Will Larson's article "Getting in the Room" (*https://oreil.ly/us7eX*) emphasizes that as well as adding value to the room, you need to reduce the *cost* of including you: show up prepared, speak concisely, and be a collaborative, low-friction contributor. If you make the room less effective at making decisions or sharing information quickly, you won't be invited back.

If you *don't* get into the room, don't take it personally, especially in orgs where people are still figuring out what their staff+ engineers are for and aren't yet on board with it being a leadership role. While they work that out, you'll have more influence (and will appear more of a leader) if you're friendly and do good work than if you grouse about not being invited to things. Understand the situation, be kind, and, as I said in Chapter 1, never be a jerk.

There are also some rooms you just shouldn't be in. If you're decidedly on the individual contributor track, you usually shouldn't be part of discussions about compensation, performance management, and other manager-track things. You might bring information to your manager or director that affects those decisions, but it's up to them to act. If big technical decisions are happening in the same room as those manager conversations, you could suggest splitting the topics into separate meetings.

Finally, remember that the room you're trying to get into may contain less power than you think. Years ago, I was shocked to discover that a group of directors didn't think their opinions carried a lot of weight; they were frustrated at not being able to influence the decisions of the *real* movers and shakers two levels up. It turned out that there was another "room" I hadn't ever thought about. There were probably others above that! Be realistic about what you're asking for access to.

The shadow org chart

So that's the formal decision making. If you understand that, you'll understand a lot about how your organization sets its opinions and decides what to do. But inevitably there's a whole lot of other influence going on, and some of it will, on the surface, make *no sense whatsoever*. Informal decision making doesn't follow rules based on hierarchy or job title. Those things certainly carry weight, but there's more going on.

While it's important to understand who the official technical leaders are, it's just as important to understand who they listen to and how they make decisions. What happens if Jan, the director of your infrastructure organization, seems to be entirely on board with your idea, then suddenly goes cold? If you're paying attention, you'll learn that Jan's first move in any decision is to check in with Sam, who joined the team 10 years ago. Sam is not particularly senior, but if Sam thinks something is a bad idea, you'll never get Jan on board. These influence lines aren't immediately obvious when you join an organization, so a good early step is to make some friends and ask how the organization works.

In their book *Debugging Teams: Better Productivity Through Collaboration* (*https://oreil.ly/TrqIi*), Brian W. Fitzpatrick and Ben Collins-Sussman describe the "shadow org chart": the unwritten structures through which power and influence flow. The shadow org chart helps you understand who the influencers of the group are, and it's probably not the same as the actual org chart. These influencers are the people you need to convince before a change can happen.

The authors identify "connectors" who know people all across the org, and "old-timers" who, regardless of rank or title, wield influence just from being around a long time. These folks are likely to have a good pulse on what can and can't work, and the people who do have rank and title will likely trust them and rely on their good judgment when making decisions. If you can get their buy-in, you're making good progress.

KEEPING YOUR TOPOGRAPHIC MAP UP TO DATE

I talked earlier about how important it is to keep your locator map up to date. Keeping your topographic map fresh is even more important. The facts on the ground will change quickly, and things that you think you know will stop being true. On an average day, you might need to know that:

- A team you depend on has a new lead.
- A project you've been waiting for isn't happening after all.
- Quarterly planning is about to start.
- A useful new platform is launching.
- Your product manager is about to go on extended leave.

There's a lot of information to keep up with. But you need to know it all, so you need to know what to look for. Here are some ways you can stay up to date:

Automated announcement lists and channels
Dedicated channels for sharing new design documents, announcing outages, or linking change-management tickets give everyone an easy high-level view of what's going on. If these kinds of channels don't exist and you'd find them useful, consider creating them.

Walking the floor
The Lean manufacturing (*https://oreil.ly/zfdHj*) folks talk about *gemba*, the idea of walking the manufacturing floor and seeing how things actually operate. Find some avenues to stay attached to the work that teams around you are doing. This could take the form of pairing on occasional changes, managing incidents, or doing a deploy for a system you want to know more about. Drifting too far from the technology doesn't just reduce your context; it can reduce your technical credibility. (More on that in Chapter 4.)

Lurking
I asked on Rands Leadership Slack (*https://oreil.ly/O4bad*) about how everyone approaches knowing things, and a common thread was paying attention to information that isn't *secret* exactly, but isn't necessarily for you. This included reading senior people's calendars, skimming agendas or notes for meetings you're not in, and—something that had never occurred to me—looking at the full list of Slack channels sorted by most recently created so you can see what new projects are happening.

Making time for reading

In companies with a mature documentation culture, plans and changes will often be accompanied by RFCs, design documents, product briefs, and so on. Skim for some basic context, or schedule time on your calendar to read deeply.

Checking in with your leadership

You need allies and sponsors who will tell you things. Check in often enough to hear behind-the-scenes updates and to make sure the way you're thinking is still aligned with the way your leaders are.

Talking with people

Stepping out for coffee and a chat isn't just pleasant relationship-building —it's a great source of context. If you really want perspective, talk to people outside engineering: product, sales, marketing, legal, and so on. If you're creating a product, befriend your customer support folks: they know more about what you've created than you do. Befriend the admin staff, too. Admins are smart, resourceful, and well connected. They know what's going on, and they tend to be the most fascinating people in the company. Go make friends.

"I Don't Know How to Talk to People"

Many engineers have an aversion to anything that smells like "network-ing." It makes us think of smarmy '80s power lunches. (Or is that just me?) But networking doesn't have to be cynical or grubby. If you get to know people and are friendly, sharing information and helping each other will follow.

If you're struggling to begin a conversation with someone, an easy starting point is to ask a question, take an interest in what they work on, or (genuinely) compliment something you admire about them.[7] Most people are interested in talking about their work or their priorities, and most will be happy to explain how something they're interested in works. Small talk is a learnable skill that will pay dividends throughout your

---

7 Compliment something you admire, but please don't tell your coworkers they're attractive! In general, only compliment something that the person made a choice to do. A well-written RFC, a smoothly run meeting, or a cool desk toy are all fair game.

career. (And if you're talking with someone more junior than you, it's kind of your responsibility to make it not awkward.)

IF THE TERRAIN IS STILL DIFFICULT TO NAVIGATE, BE A BRIDGE

The problems that slow down tech organizations are most often human ones: teams that don't know how to talk to each other, decisions that nobody feels empowered to make, and power struggles. These are difficult problems! As you add information to your topographical map, you may find places where it's tempting to scrawl "There be dragons" and vow to steer elsewhere. But a staff engineer can often have the most impact by going where everyone else fears to tread and making the dangerous territory easier for everyone else.

The Westrum model highlights the importance of "bridging" (see Figure 2-11), making connections between parts of the organization that otherwise would have enormous information gaps. The more you know the terrain, the easier it will be to bridge gaps by sending the email summary nobody is sending, introducing two people who should have spoken a month ago, or writing a document to show how projects connect to each other.

Figure 2-11. When quarterly planning is a long way off, staff engineers can build connections to bridge the gap between two orgs.

Google's DevOps site (*https://oreil.ly/DU4Nc*) suggests preemptively building bridges: "Identify someone in the organization whose work you don't understand (or whose work frustrates you, like procurement) and invite them to coffee or lunch."

When you can, define the scope of your job so that it crosses the tectonic plates and encompasses *all* of some system or problem domain, not just the part belonging to a single team. That way, you can catch work that is getting dropped, mediate conflicts, and help create a single story about what's happening. When there are major changes proposed, you'll have enough context to say "Yes, this migration is a good idea," or "No, we have work to do."

The Treasure Map: Remind Me Where We're Going?

We've drawn two maps so far. The *locator map* shows where we are. The *topographic map* shows how we can navigate across the organization. But where are we going? That's the purpose of our third map (see Figure 2-12).

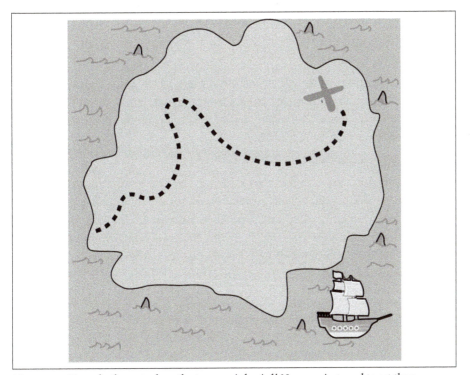

Figure 2-12. X marks the spot where the treasure is buried! Now you just need to get there.

The *treasure map* gives us a compelling story of where we're going and why we want to get there. Let's go on an adventure!

CHASING SHINY THINGS

I talked earlier in this chapter about how you need to look past your group's local problems and keep perspective about the world around you. You need that same perspective across *time*. It's easy to overfocus on short-term goals like the current feature release or the latest unhappy App Store review. But think bigger. Where are you trying to get to? Why are you doing any of this? To be clear, I'm not saying you *shouldn't* look for short-term wins. But thinking *only* about short-term goals can be limiting. If you're only thinking short term:

It'll be harder to keep everyone going in the same direction
> If the team doesn't know the big plan, either they'll go to the wrong place, or every decision will be long, complicated, and full of discussion. Sometimes navigating around difficulties will mean taking an indirect path. Everyone should be very clear about the course correction that will need to come after that milestone.

You won't finish big things
> If your team keeps focusing on short-term projects to solve local problems and pain points, you won't be able to solve bigger, long-term problems that take multiple steps. The value of your existing projects might not be clear to people outside the team either.

You'll accumulate cruft
> If teams don't know where they're going, they have two options. They can try to be flexible enough to support every future state, creating solutions that are overcomplicated and hard to maintain. Or they can make local decisions, taking the risk that their direction won't match everyone else's and that their solution will be a weird edge case that everyone else has to work around.

You'll have competing initiatives
> In an organization that relies on grassroots or bottom-up initiatives, there might be multiple people trying to rally enthusiasm around completely different directions. They're all trying to do the right thing and get people aligned, but the end result is chaos.

Engineers stop growing

Focusing only on short-term goals limits the way you think about and frame your work, and how much ownership you take of the work that falls into the cracks between tasks. If the team is trying to achieve a big project, they'll have to identify the gaps between the assigned tasks and figure out how to fill them, building skills in the process. A team that's used to iterating on short, clearly specified goals won't build muscle for bigger, more difficult projects and won't be able to tell the story of why they did what they did.

TAKING A LONGER VIEW

If everyone knows where they're going, life gets easier. There's no need to keep tight alignment along the way. Each team can be more creative in figuring out their own route, with their own narrative for the problems they'll need to solve to get there. They're less likely to go down wrong paths, and they'll have enough information to make decisions, reducing the amount of hedging and technical debt they need to incur. They can celebrate the wins along the way, while remembering that there is a long-term goal and that the real celebration won't happen until they get there.

Why are you doing whatever you're doing?

An analogy I use a lot is the technology tree that you see in many strategy games, such as *Civilization*.[8] In case you haven't, uh, *invested* way too many hours of your life on this excellent game, I'll explain how it works. You play as the ruler of a civilization, trying to build an empire. Your path to greatness includes amassing scientific knowledge, so as you go along you can choose to research various technologies. The set of available technologies form a directed graph (see Figure 2-13). At the beginning you might research, say, pottery and hunting, but as you go through the game, your skills will build on each other. In *Civilization*, you can't build a railroad without researching bridge building and steam engines. And you can't build steam engines without physics and engineering. So there's going to be a point in the game when you're researching physics but your *actual goal* is to build a railroad. You won't have the real win until you've built the bridges,

---

8 *Civilization*, now in its sixth edition, is a strategy game that has been around for decades. It uses a fog of war, and you have to make good decisions between long-term and short-term investments. I recommend playing *Civilization* to understand all things about staff engineering. Tell your boss it's research.

researched the steam engines, and ordered little hats for your train conductors. (That last bit doesn't really happen, unfortunately.) Unless you remember where you intended to go and keep working on it, you don't ever get to ride the train.

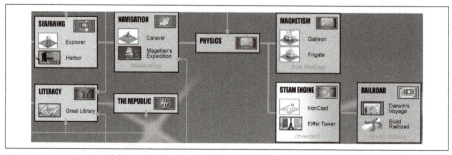

Figure 2-13. A section of the technologies graph from Sid Meier's Civilization II, created by Microprose/Activision (source: http://www.civfanatics.com).

When you're choosing a technology to invest in, often that's because it's an unavoidable step on the path to something else. You're not building a new service mesh for the joy of building a service mesh; you're building it to make your microservices framework easier to use, because you want to make it easy for new services to get set up quickly, because you want teams to be able to ship features faster. The real goal is to reduce time to market. When you know the real goal, you can step back and evaluate whether any proposed work will get you closer to it.

Sharing the map

It may take you time to dispel the fog of war and uncover the true destination of your journey. Once you do understand it, don't keep it to yourself. That means telling the story to other people and letting them understand why it matters. Your story should show where you are, where you're going, and why you're taking the steps that you are along the way. If there are sea monsters or shortcuts to know, you'll probably want those marked—but don't include any distractions. Make it easy to see what's going on. The map should describe the treasure—that is, give a clear definition of success—so that everyone knows what they're aiming for.

It's motivational to see that you're making progress, but it's also surprisingly easy to forget where you were and maybe even feel like you're getting nowhere. As the person with a map (see Figure 2-14), you're well positioned to show that everyone is getting closer to the goals (or course-correct if not). Tell the story of where you came from as well as where you're going.

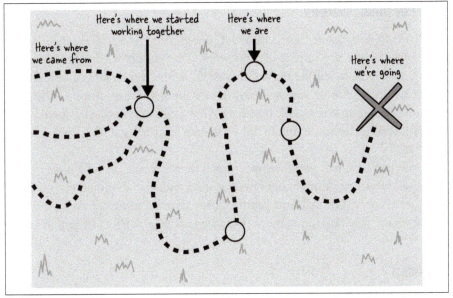

Figure 2-14. Tell the story of where you were and how much progress you've made as well as where you're going.

IF THE TREASURE MAP IS STILL UNCLEAR, IT MIGHT BE TIME TO DRAW A NEW ONE

If everyone's working from the same treasure map, your job here is done. But if you discover that there are multiple competing paths or no plans at all, you might need to help the group choose a destination. Sometimes all you need here is to write up a short summary of where you see confusion or misalignment. By spelling the facts out and sharing them, you're forcing the conversation (or, perhaps, the argument) into the open. But, after asking all of the questions, tracing the *Civilization* tech tree, encouraging the people who disagree to talk with each other, and thinking really hard, you might still conclude that no one has actually chosen a long-term destination yet, or that there are multiple competing destinations.

In that case, there's nothing more to be gained from clearing the fog of war from the map: it's time to create a new map. That's what the next chapter is all about.

Your Personal Journey

Before I close this chapter, let's talk about *your* journey. As a staff engineer, it can take longer to see the impact of your work. That means it's harder—but also more important—to tell the *story* of that work. When you look back, you should have a narrative of what you were trying to achieve and how it went. What did you and your group accomplish together? When you look ahead, you should have a story too: what are you trying to do? How does your current work contribute to that goal?

Once you have that narrative, even small tasks become part of a bigger story. Any given week's work might not be elucidating, but what did you do this month, this quarter, this year? Are you getting closer to some treasure? To understand your journey, you'll need one more map, the *trail map*. We'll draw that in Chapter 9.

To Recap

- Practice the skills of intentionally looking for a bigger picture and seeing what's happening.
- Understand your work in context: know your customers, talk with peers outside your group, understand your success metrics, and be clear on what's actually important.
- Know how your organization works and how decisions get made within it.
- Build or discover paths to allow information you need to come to you.
- Be clear about what goals everyone is aiming for.
- Think about your own work and what your journey is.

3

Creating the Big Picture

Here's my favorite story about a group that was missing a big picture. An organization I was in had an upcoming all-hands meeting. People could propose topics in advance, and there had been a question about a critical system—let's call it SystemX—that had caused some outages. I'd been tagged to respond. While I prepared my talking points, I got three almost simultaneous DMs:

- The first: "Please reassure everyone that we know SystemX has been a problem, but we're staffing up the team that supports it and adding replicas to help it scale. We don't anticipate more outages."

- At the same time: "So glad someone asked that question! We should emphasize that SystemX has been deprecated and everyone should plan to move off it."

- And: "Hey, could you tell everyone that I've set up a working group to explore how to evolve SystemX. We'll announce plans next quarter. If anyone wants to join the working group, they should contact me."

The public forum would have been a great opportunity to spread awareness of any of these three very reasonable paths forward. But why were there three different plans?

At the end of Chapter 2, we finished uncovering the existing treasure map of your organization. If your group already has one of those—a single compelling, well-understood goal and a plan to get there—your big picture is complete. You can jump to Part II, where I'll talk about how to execute on big projects. But a lot of the time, staff engineers find that the goal is not clear, or that the plan is disputed. If that's the situation you're in, read on.

In this final chapter of Part I, we're going to talk about *creating* the big picture. When the path is undefined and confusing, sometimes you need to get the group to agree on a plan and create the missing map. This map often comes in the form of a *technical vision,* describing a future state you want to get to, or a *technical strategy,* outlining how you plan to navigate challenges and achieve specific goals. I'll open by describing both of these documents, including why you'd want each one, what shapes they might take, and what kinds of things you might include in them. Then we'll look at how to work as a group to create this kind of document. We'll look at the three phases of creating the document: the approach, the writing, and the launch.[1] Finally, we'll work through a fictional case study to see some of those techniques in action. Let's lay that scenario out now, so you can start thinking about how *you'd* approach it.

The Scenario: SockMatcher Needs a Plan

SockMatcher formed a few years ago as a two-person startup aiming to solve an important problem: odd socks. People who have lost a sock upload an image or video using the company's mobile app, and a sophisticated machine learning algorithm on the backend attempts to find another user who has lost one of an identical pair. If one of the two sock owners wants to sell their odd sock to the other, the algorithm then suggests a price. Every change in sock ownership is tracked in a distributed sockchain ledger.

As you might imagine, venture capitalists went wild for it.[2] SockMatcher quickly grew into the largest odd-sock marketplace on the internet. The company has expanded to add partnerships with several bespoke sock manufacturers, personalized sock recommendations, and even gloves and buttons. It launched an external API that third parties can use for sock analysis as a platform (SaaaP). Customers love the new features.

The company's architecture has grown organically. It's all built around a single central database and data model, with a monolithic binary managing login, account subscriptions, billing, matching, personalization, image and video uploads, and so on. Product-specific logic is built into each of these functions. For instance, the billing code includes logic for how customers are charged: per successful match for socks, but as a quarterly subscription for buttons. Sock data

---

1 We're focusing on vision and strategy, but these techniques can work for any big group decisions: engineering values, coding standards, cross-organization project plans, etc.

2 If anyone wants to talk seed funding, drop me a line.

and customer data are stored in the same large datastore, which includes sensitive personally identifiable information (PII) about customers, like their names, credit card numbers, and shoe sizes.

For competitive reasons, SockMatcher has prioritized getting new features into the apps quickly, rather than building in a scalable or reusable fashion. For example, the team implemented the gloves feature as a special case of the existing sock-matching functionality, adding a field to the sock data model to allow marking an item as "left" or "right." When a user uploads an image of a glove, the software generates a mirror image, then treats the glove as just another kind of sock.

When the company decided to add button matching, several senior engineers argued that it was time to rearchitect and create a modular system where it would be easy to add new types of matchable objects. Business pressures won out, though, and button matching was also implemented as a special case of socks, with new fields in the data model to allow specifying the number of buttons in a set and the number of holes per button. The billing code, personalization subsystem, and other components contain hardcoded custom logic to handle differences in socks, gloves, and buttons, mostly implemented as *if* statements scattered throughout the codebase.

Now there's a new proposed business goal: the company wants to expand to match food storage containers and lids. This product will have different characteristics from existing ones. Unlike socks, containers and their lids aren't identical. The team will need different matching models and logic and a whole new set of vendors and partnerships so that they can offer the customer a brand-new replacement lid or container when no match is available. The company's most recent product strategy deck speculates about adding earrings, jigsaw-puzzle pieces, hubcaps, and more in the future.

The new food storage container team is ready to start scoping out the feature. They're not eager to begin working in the existing monolith: they really want to build their own independent matching microservice with their own datastore. But even if they do, they'll need code for authentication, billing, personalization, safely handling PII, and other shared functionality—all of which are currently optimized for the sock model. If they want to work autonomously, they'll need to expose this functionality from the monolith or reimplement it. Either will take time, so they anticipate some pressure to declare food storage containers to be a kind of sock and to work within the existing code, adding more edge cases alongside gloves and buttons where needed. The team is split on what the right next step is.

There are some other challenges:

- The API that was shared with third parties isn't versioned, and so it's difficult to change it; with new integrations planned, this problem will get more difficult to solve the longer it's left.

- The homegrown login functionality has always been, to quote the engineer who built it three years ago, "kind of janky." It's got a few years of growth left in it, but it's not code anyone is proud of.

- The matching functionality is the best on the market and makes customers happy, but there are times when it fails to find a match even though one is available.

- One team member has an idea for a new algorithm and system that will find matches in a fraction of the current time. They're really excited about it.

- The team responsible for operating the monolith hasn't been able to keep up with its growth and is reacting constantly to scaling problems. They're paged several times every day for full disks, failed deploys, and software bugs.

- With more and more engineers working in the same codebase and reusing existing functionality, there's more unexpected behavior, and user-visible bugs are being pushed more often. Almost every team and user is affected by almost every outage.

- Celebrities and influencers selling their socks have caused 100x spikes in user traffic, leading to complete outages. The food storage container launch might attract celebrity chefs to the platform, further increasing demand.

- Every new piece of functionality slows the monolith's build time, and unowned, flaky tests aren't helping: it typically takes three hours to build and deploy a new version, lengthening most incidents.

- App Store reviews for the mobile app have begun to trend downward; many of the one-star reviews note that availability has been poor.

Although this scenario includes a lot of problems, many of them have straightforward technical solutions. Every new person who joins the company suggests changes: sharding the datastores, versioning the APIs, extracting functionality from the monolith, and so on. Various working groups have kicked off. They always start as a room of 20 people who care deeply but don't agree, then get mired in no one having time to focus on them. After all, there's feature work to do that feels more important or more likely to succeed. The engineering organization can't seem to get momentum behind any single initiative.

What would you do? We'll return to this scenario at the end of the chapter.

What's a Vision? What's a Strategy?

Should you build new functionality as a reusable platform or as part of a specific product? Should teams learn the difficult new framework or stick with the popular deprecated one? Maybe each team can make their own decision: decentralized decision making can let organizations move more quickly and solve their own problems. But there can be disadvantages when each team decides for itself:

- There can be a "tragedy of the commons" (*https://oreil.ly/KEPbi*), where teams pursuing their own best action without coordinating with others leads to an outcome that is bad for everyone. There's that local maximum again.

- Shared concerns can be neglected because no one group has the authority or an incentive to fix them alone.

- Teams may be missing enough context to make the best decision. The people taking action might be different from the people experiencing the outcomes, or might be separated in time from them.

When there are major unmade decisions, projects get slowed or blocked. Nobody wants to delay their project for the sake of a long and painful fight to make a controversial decision or choose a standard. Instead, groups make locally good decisions that solve their own immediate problems. Each team chooses directions based on their own preferences or on rumors about the organization's technical direction—and often embed those choices in the solution they create.

Postponing the big underlying questions makes them even harder to solve down the line.[3]

When organizations realize they need to solve some of these big underlying problems, the words *vision* and *strategy* get thrown around a lot. You'll hear these used both interchangeably and to mean distinct things. To avoid confusion over terminology, let's start with some working definitions we can use throughout this chapter.

WHAT'S A TECHNICAL VISION?

A *technical vision* describes the future as you'd like it to be once the objectives have been achieved and the biggest problems are solved. Describing how everything will be *after* the work is done makes it easier for everyone to imagine that world without getting hung up on the details of getting there.

You can write a technical vision at any scope, from a grand picture of the whole engineering organization down to a single team's work. Your vision may inherit from documents at larger scopes, and it may influence smaller ones (see Figure 3-1 for some examples).

A vision creates a shared reality. As a staff engineer who can see the big picture, you can probably imagine a better state for your architecture, code, processes, technology, and teams. The problem is, many of the other senior people around you probably can too, and your ideas might not all line up. Even if you think you all agree, it's easy to make assumptions or gloss over details, missing big differences of opinion until they cause conflict. The tremendous power of the written word makes it much harder to misunderstand one another.

---

3 You might notice that these kinds of topics still consume a ton of discussion time in code and design reviews, even though they're not at the core of any particular change. If that's happening, or if you're seeing designs that have *contradictory* baked-in assumptions, that's a sign that you need to make some big central decisions, independent of any particular project or launch.

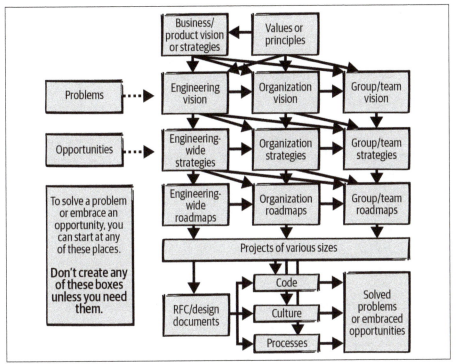

Figure 3-1. Depending on the size of the problem, you might start with an engineering-wide vision, a team-scoped vision, or something in between. Don't create a vision, strategy, etc., unless you need it.

A technical vision is sometimes called a "north star" or "shining city on the hill." It doesn't set out to make *all* of the decisions, but it should remove sources of conflict or ambiguity and empower everyone to choose their own path while being confident that they'll end up at the right place.

Resources for Writing a Technical Vision

If you're setting out to write a technology vision document, here are some resources I recommend:

- *Fundamentals of Software Architecture* by Mark Richards and Neal Ford (O'Reilly)
- Chapter 4 of *Making Things Happen* by Scott Berkun (O'Reilly)

- "How to Set the Technical Direction for Your Team" (*https:// oreil.ly/zhD2Q*), by James Hood of Amazon
- "Writing Our 3-Year Technical Vision" (*https://oreil.ly/Rew44*), by Daniel Micol of Eventbrite

---

There's no particular standard for what a vision looks like. It could be a pithy inspirational "vision statement" sentence, a 20-page essay, or a slide deck. It might include:

- A description of high-level values and goals
- A set of principles to use for making decisions
- A summary of decisions that have already been made
- An architectural diagram

It could be very detailed and go into technology choices, or it could stay high-level and leave all of the details to whoever is implementing it.

Whatever you create, it should be clear and opinionated, it should describe a realistic better future, and it should meet your organization's needs. If you could wave a magic wand and be done, what would your architecture, processes, teams, culture, or capabilities be? That's your vision.

WHAT'S A TECHNICAL STRATEGY?

A *strategy* is a plan of action. It's how you intend to achieve your goals, navigating past the obstacles you'll meet along the way. That means understanding where you want to go (this could be the vision we just discussed!) as well as the challenges in your path. When I use the word *strategy* in this chapter, I always mean a specific document, not just being a strategic sort of thinker.

A technology strategy might underpin a business or product strategy. It might be a partner document for a technical vision, or it might tackle a subset of that vision, perhaps for one of the organizations, products, or technology areas it encompasses. Or it might stand entirely alone.

Just like a technical vision, a technical strategy should bring clarity—not about the destination, but about the path there. It should address specific challenges in realistic ways, provide strong direction, and define actions that the group should prioritize along the way. A strategy won't make all of the decisions, but it should have enough information to overcome whatever difficulties are stopping the group from getting to where it needs to go.

Resources for Writing a Technical Strategy

The canonical book on strategy is *Good Strategy/Bad Strategy* by Richard Rumelt (Currency). I recommend taking the time to read it if you can. Other great resources include:

- "Technical Strategy Power Chords" (*https://oreil.ly/EDODP*) by Patrick Shields
- "Getting to Commitment: Tackling Broad Technical Problems in Large Organizations" (*https://oreil.ly/RKUwO*) by Mattie Toia
- "A Survey of Engineering Strategies" (*https://oreil.ly/tF2TU*) by Will Larson
- *Technology Strategy Patterns* by Eben Hewitt (O'Reilly)
- Rands Leadership Slack (*https://oreil.ly/O4bad*), specifically the channels #technical-strategy and #books-good-strategy-bad-strategy

---

Just like a technology vision, a strategy could be a page or two or it could be a 60-page behemoth. It will likely include a diagnosis of the current state of the world, including specific challenges to be overcome, and a clear path forward for addressing those challenges. It might include a prioritized list of projects that should be tackled, perhaps with success criteria for those projects. Depending on its scope, it could include broad, high-level direction or decisions on a specific set of difficult choices, explaining the trade-offs for each one.

In *Good Strategy/Bad Strategy*, Rumelt describes "the kernel of a strategy": a diagnosis of the problems, a guiding policy, and actions that will bypass the challenges. Let's look at each one.

The diagnosis

"What's going on here?" The diagnosis of your situation needs to be simpler than the messy reality, perhaps by finding patterns in the noise or using metaphors or mental models to make the problem easy to understand. You're trying to distill the situation you're in down to its most essential characteristics so that it's possible to really comprehend it. This is *difficult*. It will take time.

Guiding policy

The guiding policy is your approach to bypassing the obstacles described in the diagnosis. It should give you a clear direction and make the decisions that follow easier. Rumelt says it should be short and clear, "a signpost, marking the direction forward."

Coherent actions

Once you've got a diagnosis and guiding policy, you can get specific about the actions you're going to take—and the ones you won't. Your actions will almost certainly involve more than technology: you might have organizational changes, new processes or teams, or changes to projects. I really can't stress this enough: you'll commit time and people to *these* actions rather than to the long list of other ideas that were on the table at the start. This kind of focus will likely mean that you and others don't get to do some things you've been excited about. It is what it is.

A strategy should draw on your advantages. For example, when Isaac Perez Moncho, an engineering manager in London, was writing an engineering strategy for a company, he looked to create positive feedback loops. That company's product engineering teams were facing many problems, he told me: lack of tooling, too many incidents, and poor deployments. But he had an advantage: a great DevOps team who could solve those problems, if only they had more time. His guiding policy was freeing up some of the DevOps team's time. Making space for them let them automate processes and free up even more time, creating a positive feedback loop that made them available to solve other problems. Think about ways to amplify your advantages in a self-reinforcing way.

Finally, a strategy isn't an aspirational description of what someone else would do in a perfect world. It has to be realistic and acknowledge the constraints of your situation. A strategy that can't get staffed in your organization is a waste of your time.

DO YOU REALLY NEED VISION AND STRATEGY DOCUMENTS?

Technical visions and strategies bring clarity, but they can be overkill for a lot of situations. Don't make unnecessary work for yourself. If it's easy to describe the state you're trying to get to or the problem you're trying to solve, what you actually want might be more like the goals section of a design document, or even the description of a pull request.[4] If everyone can get aligned without the document, you probably don't need it.[5]

If you're sure you need something, think about what shape it should take. Adapt to what your organization needs and will support. For example, if a lack of direction is slowing you all down, you might want to get a group together to create an abstract high-level vision, then get more concrete about how to implement it. If you're preparing for company growth, your CTO might ask you to get a group together from across engineering and describe what your architecture and processes will look like in three years. But if your group is repeatedly getting stuck on a particular missing architectural decision, don't spend too much time on philosophy: make a call on the specific item that's blocking you.

Writing technical vision or strategy takes time. If you can achieve the same outcome in a more lightweight way, do that instead. Create what your organization needs and no more.

The Approach

Creating a vision, a strategy, or any other form of cross-team document is a big project. There will be a ton of preparation, then a ton of iteration and alignment. Bear in mind that getting people to agree isn't a chore that stands in between you and the real work of solving the problem: the agreement *is* the work. Any insight or bold vision you're bringing to the project is only going to be worth anything if you can bring people along on the journey with you. Just like you wouldn't admire time spent on an engineering solution that ignores the laws of physics, advocating for a project that you know you won't be able to convince your organization to do isn't a good use of your time.

Although I've talked about strategies and visions separately up until now, for the rest of this chapter I'm not going to distinguish between them much. They're

---

4 I'll talk about these kinds of documents when we're looking at project execution in Chapter 5.

5 That said, Patrick Shields, a staff engineer at Stripe, told me once that he encourages people to write small strategies for all sorts of things because "you need to learn to play amateur basketball before you drop into the NBA." It's an excellent point.

very different things, but both involve getting people together, making decisions, bringing your organization along, and telling a story. You can use the same approach for creating other big-picture documents, like a technology radar (*https://oreil.ly/lPzPZ*) or a set of engineering values.

Creating any of these documents is a classic "1 percent inspiration and 99 percent perspiration" endeavor, but if you prepare properly, you increase your chances of launching something that actually gets used. I'll talk about some of the prep work that can set you up for success, and then I'll invite you to think through the outputs of that prep work, evaluate whether you can succeed, and decide whether to make the project official.

EMBRACE THE BORING IDEAS

I don't know about you, but when I was new to the industry, I thought that very senior engineers were wizards who would spend their days coming up with insightful game-changing solutions to terrifyingly deep technical problems. I imagined it to be something like a *Star Trek: The Next Generation* episode where there's an impending warp core antimatter containment failure or what have you, and everyone's out of ideas and freaking out, but then suddenly Geordi La Forge or Wesley Crusher exclaims, "Wait! What if we <extreme technobabble>" and taps eight characters on a touch screen, and the Enterprise is saved with seconds to spare. Phew!

Real life is a bit different. OK, sometimes "What if we <extreme technobabble>" actually is the answer: especially in very small companies, sometimes you really are stuck until an experienced person drops in to describe a solution and save the day. But if there are senior people around, most likely there are already plenty of good ideas. The gap is getting everyone to agree on what to do.

As you go into this project to create a vision or a strategy, be prepared for your work to involve existing ideas, not brand-new ones. As Camille Fournier, author of *The Manager's Path*, wrote (*https://oreil.ly/WgEPL*):

> I kind of think writing about engineering strategy is hard because good strategy is pretty boring, and it's kind of boring to write about. Also I think when people hear "strategy" they think "innovation." If you write something interesting, it's probably wrong. This seems directionally correct, certainly most strategy is synthesis and that point isn't often made!

Will Larson adds (*https://oreil.ly/3TH5a*), "If you can't resist the urge to include your most brilliant ideas in the process, then you can include them in your prework. Write all of your best ideas in a giant document, delete it, and never mention any of them again. Now...your head is cleared for the work ahead."

Creating something that feels "obvious" can feel anticlimactic when you're writing it: we'd all love to show up with a genius visionary idea and save the USS *Enterprise*! But usually what's needed is someone who's willing to weigh up all of the possible solutions, make the case for what to do and not do, align everyone, and be brave enough to make the (potentially wrong!) decision.

JOIN AN EXPEDITION IN PROGRESS

If someone is already working on the kind of document you want to create, don't compete—join their journey. Here are three ways to do that:

Share the lead

You can bring leadership to an existing project without taking over. Suggest a formal split that gives each of you a chance to lead in a compelling way. One takes the overall project, for example, while the other leads some individual initiatives inside that work. You could also suggest co-leading: take it in turns to be the primary author, and split up the work as it comes along. If the leaders are enthusiastic about each other's ideas and are all pushing in the same direction, this can make for a very effective team.

Follow their lead

Put your ego aside and follow their plan. If they're less experienced than you, you can have a huge impact by nudging them in the right direction and helping make the work as good as it can be. Being the grizzled, experienced best supporting actor is an amazing role.[6] You can use your deep technical knowledge to fill gaps in their skills, for example, spelunking in a legacy codebase to understand exactly how something works. You can also advocate for the plan in rooms they aren't in. Back them up and help make the thing happen.

---

6 A caveat: if you're someone who *always* takes the back seat and cheers on others, make sure your organization recognizes that leadership. If not, make sure you have some opportunities to shine too. I'll talk about credibility and social capital in Chapter 4.

Step away

A third approach, of course, is to decide the work is going to succeed without you, be enthusiastic about it, and go find something else to do. If the project doesn't need you, find one that does.

It can be difficult to let other people lead when their direction is not where you'd planned to go. Tech companies' promotion systems can incentivize engineers to feel like they need to "win" a technical direction or be the face of a project. This competition can lead to "ape games": dominance plays and politicking, where each person tries to establish themselves as the leader in the space, considering other people's ideas to be threatening. It's toxic to collaboration and makes success much harder to achieve.

My friend Robert Konigsberg, a staff engineer and tech lead at Google, always says, "Don't forget that just because something came from *your* brain doesn't make it the best idea." If you tend to equate being right with "winning," step back and focus on the actual goal. Practice perspective: is their direction wrong, or is it just *different?* Would you advocate just as hard for the path you want if it had been a colleague's idea? Even if it's better, be wary of fighting for a marginally better path at the cost of not making a decision at all. As Will Larson writes (*https://oreil.ly/TmbEr*), "Give your support quickly to other leaders who are working to make improvements. Even if you disagree with their initial approach, someone trustworthy leading a project will almost always get to a good outcome."

What if you *don't* think the person's ideas or leadership can work, even with your support? While you sometimes need to be flexible, that shouldn't extend to endorsing ideas you think are dangerous or harmful. Even then, try to join the existing journey and change its direction, rather than setting up a competing initiative from scratch: you'll have allies in place and momentum already built, and you'll learn from whatever they've done so far.

If there really is no way for multiple people to succeed on the same project without playing ape games, consider going somewhere with more available scope (and a healthier culture).

GET A SPONSOR

Except in the most grassroots of cultures, any big effort that doesn't have high-level support is unlikely to succeed. A vision or strategy can begin without sponsorship, but turning it into reality later will be a challenge. Even early on, a sponsor helps clarify and justify the work. If your director or VP is on board with your plans from the start, then what you're creating is implicitly the *organization's* treasure map, not just yours—reducing the risk that you're wasting your time. Sponsorship can also add hierarchy to groups that would otherwise get stuck attempting consensus. The sponsor can set success criteria and act as a tie-breaker when decisions are stuck in committee. They can nominate a lead or "decider"—sometimes called a directly responsible individual, or DRI—who will get the final say when the group is stuck. You don't necessarily need that, but keep it in mind as an option that's available to you.

Getting a sponsor might not be easy. Executives are busy and you're implicitly asking them to commit resources, people, and time to your proposal over something else they'd intended to do. Maximize your chances by bringing the potential sponsor something they want. While a proposal that's good for the company is a great start, you'll get further with one that matches the director's own goals or solves a problem they care about. Find out what's blocking their goals, and see if the problem you're trying to solve lines up with those. The sponsor will also have some "objectives that are always true": if you can make their teams more productive or (genuinely) happier, that can be a compelling reason for them to support your work.

Think about and practice your "elevator pitch" before trying to convince a sponsor to get on board with your project: if you can't convince them in 50 words, you may not be able to convince them at all. I once tried to talk Melissa Binde, at the time a director of engineering at Google, into sponsoring a project I cared deeply about. I went into all sorts of detail as I tried to make it important to her, too. My spiel wasn't convincing *at all*—but Melissa kindly took the opportunity to coach: "The way you're telling this story doesn't make me care, and it won't make anyone else care either. Try again—tell the story from a different angle." She let me try a few different project rationales, and told me which ones resonated. You will almost never get an opportunity like that, so go in with your pitch already polished.

Can a staff+ engineer be a project sponsor? In my experience, no, not directly. A sponsor needs the power to decide what an organization spends time and staffing on, and such decisions are usually up to the local director or VP.

Once you have the sponsorship, check in enough to make sure you *still* have it. Sean Rees, principal engineer at Reddit, says that one of the biggest mistakes a staff+ engineer can make is not maintaining their executive sponsorship: "I think this one is pernicious because you can start sponsored and have that wane as realities change...and then have to navigate the tricky waters of getting back into alignment."

CHOOSE YOUR CORE GROUP

Some people are very disciplined about setting out on a journey and not getting distracted by side quests, but most of us benefit from a little accountability. Working with other people will give you that accountability. When one person is distracted or flagging, the others will keep the momentum going, and knowing that you'll need to check in about the work can be a powerful motivator. Working in a group also lets you pool your contacts, social capital, and credibility, so you can reach a broader swathe of the organization. You'll have to work to get everyone aligned, but since ultimately you'll need to align with your whole organization, getting your core group to agree can help you uncover potential disagreements early on.

Aim to recruit a *small* core group to help you create the document, as well as a broader group of general allies and supporters. Of course, if you've joined someone else's journey, you're going to be part of *their* core group instead. (For the sake of simplicity, the rest of this chapter assumes you're leading the effort.)

Who do you want in your group? Pull out your topographic map. Who do you need on your side? Who's going to be opposed? If there are dissenters who will hate any decision made without them, you have two options: make very sure that you have enough support to counter their naysaying, or bring them along from the start. Bringing them along will be easier if you understand *why* they're against the work and what they'd like to see happen instead.

While you may have many colleagues who care about what you're writing and want to help, keep the core group manageable: two to four people (including you) is ideal. A time commitment can help you here: if everyone who's part of the core team needs to commit 8 or 12 hours a week to this work, you'll be able to keep the group small without excluding anybody. (This is a good way to keep "tourists" out of the way, too.) Outside the core team, you can offer more lightweight involvement: you'll interview them, try to represent their point of view in your work, and let them review early drafts.

Once you have your core group, be prepared to let them work! Be clear from the start about whether you consider yourself the lead and ultimate decider of this project or more of a first among equals (*https://oreil.ly/BKNe3*). If you're later going to want to use tiebreaker powers or pull rank, highlight your role as lead from the start, perhaps by adding a "roles" section to whatever kickoff documentation you're creating. But whether you're the lead or not, let your group have ideas, drive the project forward, and talk about it without redirecting all questions to you. Offer opportunities to lead, and make sure you're supportive when they take initiative. Their momentum will help you move along faster, so don't hold them back.

As for your broader group of allies, keep them engaged: interview them and use the information they give you. Send them updates on how the document is going. Invite them to comment on early drafts. Your path will be much easier if you have a group of influential people who support what you're doing, and the knowledge they bring from across the organization will yield better results, too.

SET SCOPE

As you think about the specific problem or problems you're trying to solve, consider how much they sprawl across the organization. Do you want to influence all of engineering, a team, a set of systems? Your plan's scope may match the scope of your role as a staff engineer, but might also extend well beyond it.[7]

Aim to cover enough ground to actually solve your problem, but be conscious of your skill level and the scope of your influence. If you're trying to make a major change to, say, your networks, you'd be wise to include someone from the networks team in the core group and to build credibility with that team. Otherwise, you're setting yourself up for conflict and failure. If you're trying to write a plan for areas of the company that are well outside your sphere of influence, make sure you have a sponsor who has influence there—and ideally some other core group members who have clear maps of those parts of the organization.

Be practical about what's possible. If your vision of the future involves something entirely out of your control, like a change of CEO or an adjustment in your customers' buying patterns, you've crossed into magical thinking. Work around your fixed constraints, rather than ignoring them or wishing they were different.

That said, if you're writing a vision or strategy for just your part of the company, understand that a higher-level plan may disrupt yours. Even if you're writing something engineering-wide, a change in business direction can invalidate all of your decisions. Be prepared to revisit your vision at intervals and make sure

---

7 If you jumped ahead to this strategy chapter and skipped all of that "What even is your job?" introspection earlier in the book, your *scope* is the domain, team, or teams that you feel responsible for. It often covers the same area that your manager covers, but not always.

it's still the right fit for your organization. As you make progress on your vision or strategy, you may find that your scope changes. That's OK! Just be clear that it has.

Be clear, too, about what kind of document you intend to create. I recommend starting by having each of the core group members be really explicit about what documents, presentations, or bumper stickers they hope will exist at the end of the work. Then choose a document type and format that makes sense to you and, most importantly, to your sponsor. If they are enthusiastic about a particular approach, soul-search before doing something else. Don't make your life harder than it has to be.

MAKE SURE IT'S ACHIEVABLE

As you think through the project ahead, how many big problems do you see? Are there decisions that you really don't know how to make, or massive technical difficulties? Having one or two problems you don't know how to solve doesn't mean you shouldn't wade in, but have a think about whether the problem is solvable at all.

A practical step you can take here is to talk with someone who's done something similar before. Ask something like: "I'm writing a vision/strategy and I currently see three problems ahead that I don't yet know how to tackle. I'm willing to try, but I don't want to waste my time if this isn't solvable. Can you give me a gut check on whether I'll hit a dead end?" Or perhaps: "Everything ahead seems doable and I only have one problem, but it's that my boss thinks this is a waste of time and wants me to focus on something else entirely. Is this worth continuing?"[8]

Maybe you think the problem is important and could be solved, but not currently by *you*. Is there a coach or mentor who could help you stretch to do it? Or is this just too big for where you are in your career? If you're a staff engineer balking at a problem scoped for a principal engineer, that doesn't reflect badly on you: you're actually doing pretty good risk analysis.

---

8 The answer to this one is usually no.

If, at the end of this analysis, you decide that the problem isn't solvable, or at least not by you, you have five options:

- Lie to yourself, cross your fingers, and do it anyway.
- Recruit someone who has skills that you're missing, and either work with them or ask them to lead the project (and give you a subsection of it to hone your skills on).
- Reduce your scope, add in the fixed constraints, and start this chapter again with a differently shaped problem to solve.
- Accept that nobody's going to write a vision/strategy to solve the problems you can see, conclude that your company will probably be OK without one, and go work on something else.
- Accept that nobody's going to write a vision/strategy to solve the problems you can see, conclude that your company *won't* be OK without one, and update your resume.

MAKE IT OFFICIAL

Before we move on, let's recap those questions we've asked along the way. Here's a checklist to consider before starting to create a technical vision or strategy.

- ☐ We need this.
- ☐ I know the solution will be boring and obvious.
- ☐ There isn't an existing effort (or I've joined it).
- ☐ There's organizational support.
- ☐ We agree on what we're creating.
- ☐ The problem is solvable (by me).
- ☐ I'm not lying to myself on any of the above.

Introspect a bit on that last question. If you can't check all of these boxes, my opinion is that you shouldn't continue. There's a high opportunity cost if you spend your time on a vision or strategy instead of any of the other work that needs a staff+ engineer.

If you do feel ready to go, though, here's one final question: are you ready to commit to the work and start working on it "out loud"? This might be a good time to formally set up the vision or strategy creation as a project, with kickoff documentation, milestones, timelines, and expectations for reporting progress. If you have *any* tendency to procrastinate or get distracted, these structures are especially important.

Your level of transparency here will depend on your knowledge of your organization: think about the topographical map you made last chapter. If you can be open about work like this, it will be easier for people to bring you information and gravitate toward you to help. If you feel you need to create a vision or strategy in secret, understand why. Does it mean you don't have enough support? Are you unsure of your own level of confidence and commitment? If you need to do a bit of the work first to convince yourself that you're going to stick with it, well, I won't judge, but make it official as soon as you can. If you set everyone's expectations, you're less likely to meet with competing efforts—or at least you'll find out about them early.

The Writing

The prep work is done, the project is official, you've got a sponsor, you've got a core group, and you've chosen a document format. The work you're doing is framed and scoped. Time to start writing the document for real.

THE WRITING LOOP

In this section, I'm going to talk through some techniques for actually creating your vision, strategy, or other broad document. We'll look at writing, interviewing people, thinking, and making decisions, as well as staying aligned while you do it. These techniques won't necessarily happen in the order I'm listing them. In fact, probably you'll do most of them many times (see Figure 3-2), maybe even occasionally dropping back to steps in the "Approach" section as your perspective changes.

There will always be more information, so notice when you start to get diminishing returns from this loop. It's very easy for a vision or a strategy to keep dragging on, particularly if it's not your primary project, so timebox this work and give yourself some deadlines. If you've set up milestones, use them as a reminder to stop iterating and wrap up. Don't be afraid to stop, even if it's not 'perfect": you can—and should—revisit the document regularly to see what context has changed. If you've missed something, there'll be opportunities to add it later.

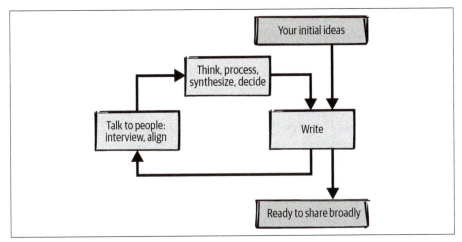

Figure 3-2. Iterating on writing a vision or strategy.

Initial ideas

Here are some questions you might ask when you're initially thinking about creating a vision or strategy. These questions are just a starting point: there will be many stakeholders and perspectives, and this is just an exercise to help you get your thoughts in order before talking to others.

What documents already exist? If there are visions or strategies that encompass yours, like company goals or values or a published product direction, you should "inherit" any constraints they've set. Here's where the perspective from your locator map will come in handy. If you're writing a wide-scale technical vision, you should know what your organization hopes to achieve in the next few years. The future that you're envisioning should include success for those existing plans and for the technical changes that have to happen to underpin them.

If there are team-level or group-level documents at a smaller scope than yours, be aware of those too. Sometimes it's inevitable that a vision or strategy with a broader scope will cause a narrower one to have to change, but understand the disruption that will cause and weigh it up when you're thinking about trade-offs.

What needs to change? What's difficult right now? If your teams are complaining about being blocked by dependencies on other teams, you might want to emphasize autonomy. If new features are slow to ship, maybe what you want is fast iteration speed. If your product is down as much as it's up, maybe you need a focus on reliability.

Knowing what you do about your company, where should your group be investing? Mark Richards and Neal Ford talk about "architectural characteristics" in software: scalability, extensibility, availability, and so on.[9] Which of those characteristics will you need to invest in as the business expands or changes?

Think big. If you're working in a codebase that takes a day to build and deploy, it might be tempting to wish for incremental improvement. "This needs to take only half a day!" But go further than that. If you set a goal of 20-minute deploys, the teams pushing toward that goal have an incentive to have bigger, braver ideas. Maybe they'll contemplate replacing the CI/CD system, or discarding a test framework that can never be compatible with that goal. Inspire people to get creative. (But, as I'll discuss later in this chapter, don't set goals that are impossible for your organization to achieve.)

What's great as it is? If you have snappy performance, rock-solid reliability, and a simple and clean UI, make sure your future state includes keeping the things that are working well. Maybe you'll end up deliberately trading off some of that greatness for something else you want more, but don't do it unconsciously.

What's important? You may be noticing a pattern here—this question has come up in every chapter so far (and it'll come up again, too). Your vision or strategy will influence the work of many senior people. Don't waste their time or yours on things that don't matter. If you're getting teams to do an expensive migration from one system to another, for example, there had better be a treasure at the end. The more effort it's going to take, the better the treasure needs to be.

What will Future You wish that Present You had done? Last one! I love the technique of envisioning a conversation with Future Me, two or three years older (and hopefully wiser). I'd ask what the world looks like, what we did, and what we wish we'd done. Which problems are getting a little worse every quarter and are going to be a real mess if ignored? Do your future self a favor if you can, and don't ignore those. I call these favors "sending a cookie into the future": it's a small but heartfelt gift from your current self to your future self.

Writing

At the end of thinking through these questions, you may be starting to identify themes and have an opinion about what's most pressing. This may be a good

---

9 They've got a thorough list in Chapter 4 of their book, *Fundamentals of Software Architecture* (O'Reilly).

time to start a rough first draft. Be prepared to let other people change your mind, though, and maybe change it a lot. You'll edit and iterate as you talk to other people and decide what to do next.

Writing as a group can be tricky. Here are two approaches:

Have the leader write a first draft for discussion

One option is that the leader of the group (again, I'm assuming that's you) writes up a first draft for discussion. This approach is great for giving the document a consistent voice and set of concerns. Be aware, though, that this draft will inevitably be influenced by its author's interests and affiliations. Reviewers and editors will be biased by what's already in it, especially if the person writing it has more influence or seniority than they do.

You can help mitigate this effect by spending a lot of time talking as a group before you start to write. If you don't feel strongly about some decisions or have even chosen arbitrarily, flag that clearly: "I rolled some dice and chose this direction. I think it's a reasonable default if we can't come to a decision. I bet we can do better, though." Make absolutely sure that anyone who has more knowledge than you on this system will feel safe disagreeing with you. "Strong ideas weakly held" only works if you're crystal clear that that's what is happening.

Aggregate multiple first drafts

The other approach is that each person in the core group writes their own first draft, and then someone aggregates it afterward.[10] Mojtaba Hosseini, a director of engineering at Zapier, told me about a group that took this approach at a previous company. Having multiple documents was a great way to get everyone's unbiased opinions, he said, but some participants ended up getting emotionally invested in their own document, criticizing the others instead of contributing to them. That group hadn't nominated anyone to combine the drafts or act as a tiebreaker when two disagreed. Hosseini advises making clear up front that these documents are all inputs to one final document that everyone gets to review at the end—no one document will be the "winner." Set expectations about who will write that final version and mediate disagreements.

---

10 Sarah Grey, development editor for this book, says that this could be an editorial nightmare if not handled carefully and that the thought of aggregating all these drafts gives her a headache. If you take this path, know what you're getting into.

Interviews

Your core group's ideas and opinions will reflect only the experiences of the people in that group. You might not know what you don't know about what's difficult in teams other than yours—so put your preconceptions aside and talk to people. Lots of people. Don't just pick the colleagues you already know and like: chances are they're organizationally pretty close to you. Seek out the leaders, the influencers, and the people close to the work in other areas.

Early on, you might ask broad, open-ended questions in your interviews.

- "We're creating a plan for X. What's important to include?"
- What's it like working with…?"
- If you could wave a magic wand, what would be different about…?"

When you have scoped and framed your work, you might scope the conversation by describing how you're thinking about the topic, sharing a work-in-progress document, or asking for their reaction to a straw man approach. Optimize for getting as much useful information as possible and for making your interviewee feel like part of what you're doing. I always end this kind of interview with "What else should I have asked you? Is there anything important I missed?"

Interviewing has another benefit: it shows the interviewees that you value their ideas and intend to include them in what you write. You'll hear about problems that you hadn't considered and new opinions about problems you already know about. Other people may disagree with you about what the biggest problems are. Have an open mind and take their thoughts seriously.

Thinking time

However you and your group like to process information (whiteboarding, writing, drawing diagrams, structured debate, sitting in silence and staring at a wall), make sure you give yourselves a lot of time to do that. I think best by writing, so when working on a vision or strategy, I need to write out my thoughts, then refine and edit them for a long time until they make more sense to me. I also get a lot from just talking through the ideas with colleagues, asking ourselves questions and trying to pick apart nuances. My colleague Carl, by contrast, likes to load up his brain with information and sleep on it: he'll usually have new insights the next morning. In some cases, you'll be able to build prototypes to test out your ideas. In others, the strategy will be more high-level and you'll have to walk through the consequences as a thought experiment instead.

Be open to shifts in your thinking. As you make progress in identifying the areas of focus or the challenges to be solved, you'll notice that you're finding new ways to talk about them. Lean into this and help it happen. The mental models and abstractions you build will help you think bigger thoughts.

Thinking time is also a good time to check in on your motivations. Notice if you're describing a problem in terms of a solution you've already chosen—this can be a mental block for a lot of engineers. We start out by comparing problems to solve, but find ourselves talking in terms of technology or architecture we "should" be using that would make everything better. As Cindy Sridharan says (*https://oreil.ly/DggW1*), "A charismatic or persuasive engineer can successfully sell their passion project as one that's beneficial to the org, when the benefits might at best be superficial." Be especially aware of what you're selling to yourself! When looking at the work you're proposing to do, ask "Yes, but *why?*" over and over again until you're *sure* the explanation maps back to the goal you're trying to achieve. If the connection is tenuous, be honest about that. Your pet project's time will come. This isn't it.

MAKE DECISIONS

At every stage of creating a vision or a strategy, you're going to need to make decisions. This chapter has discussed many early decisions: what kind of document to create, who to get involved, who to ask for sponsorship, how to scope your ambition, which goals or problems to focus on, who to interview, and how to frame the work. As you work through your vision or strategy, you'll need to weigh trade-offs, decide how to solve a problem, and decide which group of people won't get their wish.

Trade-offs

The reason decisions are hard is that all options have pros and cons. No matter what you choose, there will be disadvantages. Weighing your priorities in advance can help you decide which disadvantages you're willing to accept. The same holds for advantages: every solution is probably good for something!

One of the best ways I've seen to clarify trade-offs is to compare two positive attributes of the outcome you want. I've heard these called *even over* statements. When you say "We will optimize for ease of use *even over* performance" or "We will choose long-term support *even over* time to market," you're saying, (to quote

"The Agile Manifesto" (*https://agilemanifesto.org*)), "While there is value in the items on the right, we value the items on the left more." This framing makes the trade-offs clear.

Building consensus

Sometimes none of the options on the table can make everyone happy. Do try to get aligned, but don't block on full consensus: you might be waiting forever, and so you're back to implicitly choosing the status quo. Take a tactic from the Internet Engineering Task Force (IETF), whose principles (*https://oreil.ly/x3Bds*) of decision making famously reject "kings, presidents, and voting" in favor of what they call "rough consensus," positing that "lack of disagreement is more important than agreement." In other words, take the sense of the group, but don't insist that everyone must perfectly agree. Rather than asking, "Is everyone OK with choice A?" ask, "Can anyone not live with choice A?"

When IETF working groups make decisions, they're looking for a large majority of the group to agree and for the major dissenting points to have been addressed and debated, even if not to everyone's satisfaction. There may not be an outcome that makes everyone happy, and they're OK with that. Mark Nottingham's foreword to *Learning HTTP/2* by Stephen Ludin and Javier Garza (O'Reilly) notes of his experience in one such group: "In a few cases it was agreed that moving forward was more important than one person's argument carrying the day, so we made decisions by flipping a coin."

If rough consensus can't get you to a conclusion and you're not ready to flip a coin, someone will need to make the call. This is why it's best to decide up front if you have, or someone else has, clear leadership authority and can act as a tiebreaker. If there's nobody in that role, you could ask your sponsor to adjudicate, but make this a last resort. Your sponsor is likely so far away from the decision that they don't have all of the context, you'd be asking for a lot of their time, and they might end up picking a path that nobody is happy with.

Not deciding is a decision (just usually not a good one)

When you're choosing between Option A and Option B, there's an implicit third option, C: *don't decide*. People often default to Option C, because it lets them stay on the fence and not upset anyone. This is the *worst* thing you can do. Decisions constrain possibilities and make it possible to make progress. Not deciding is in itself a decision to maintain the status quo, as well as the uncertainty that surrounds it. While keeping your options open indefinitely might feel like it's giving you flexibility, in the longer term your solution space stays large. Other decisions that depend on this one have to hedge to prepare for any of the possible directions you might end up choosing later.

If you realize you'll need more information to make an informed decision, what extra information do you expect to get, and how? If you choose to wait, what are you waiting *for*? Remember that you usually don't need to make the *best* decision, just a *good enough* decision. If you're stuck, timebox it. Instead of saying, "We'll choose the best storage system on the planet," try, "Let's research storage systems for the next two weeks and choose by the end of that time."

There are sometimes good arguments for postponing a decision. The excellent decision-making website The Decider (*https://thedecider.app*) lists a few: when the time and energy you would need to invest in deciding just isn't worth any of the benefits the decision will give you, when getting it wrong carries heavy penalties and getting it right carries little reward, or when you suspect the situation might just go away on its own. But the key is that you *decide* not to decide. You add "make no decision (status quo)" to your list of options and deliberately choose it. You don't just throw up your hands and walk away.

When you can't be sure that your decision is a great one, think about what could go wrong. Can you include ways to course-correct or mitigate any negative outcomes? Make it so that if you're wrong, it's not a terrible thing.

Show your work

However you make the decision, document it, including the trade-offs you considered and how you got to the decision in the end. Don't elide the disadvantages: be very clear about them and explain why you've decided this is still the right path. In some cases, it's genuinely going to be impossible to make everyone happy, but you can at least show that you've understood and considered all of the arguments. Not only is it respectful to give your coworkers this information, it also reduces the risk of having to relitigate the decision every time a new person joins the project and assumes you haven't noticed the disadvantages they can see.

Making someone unhappy is, unfortunately, inevitable when you're creating a strategy or vision. If everyone gets their wish, it's unlikely that you made any real decisions. Be empathetic and try to solve everyone's problems when you can, but make a decisive call and show why you chose what you chose.

GET ALIGNED AND STAY ALIGNED

Understand who your final audience will be. Will you need to convince a small number of fellow developers? The whole company? People outside your company? Think about how you can bring each group along on the journey with you.

Keep your sponsor up to date on what you're planning and how it's going. That doesn't mean you should send them an unedited 20-page document while you're still trying to figure out what point you're trying to make. Take the time to get your thoughts together so that you can bring them a summary of how you're approaching various problems and what your options are. Unless they want to see the work in progress, share the highlights of what you're writing, rather than the gory details. In particular, if you're writing a strategy, make sure you're aligned *at least* at the major checkpoints: after you've framed the diagnosis, after you've chosen a guiding policy, and again after you've proposed some actions. If your sponsor believes you're on the wrong path, you'll want to find out before you spend a lot more time on it.

Be reasonable

Your path should be aspirational, but not impossible. Some changes are much too expensive to justify the investment. Other efforts just won't win support in your organization. There's a political science concept called the Overton window (*https://oreil.ly/wTFcR*) (Figure 3-3): the range of ideas that are tolerated in public discourse without seeming too extreme, foolish, or risky. If your ideas are too futuristic for the people you need to believe in them, your colleagues will dismiss your document and you'll lose credibility (more on that in Chapter 4). Be aware of what your organization will accept, and don't take on an unwinnable battle.

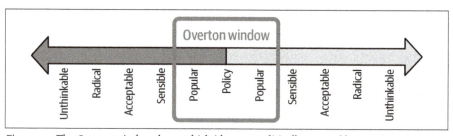

Figure 3-3. The Overton window shows which ideas are politically acceptable at a given time (source: based on an image from the Toronto Guardian (https://oreil.ly/dANiZ)).

Nemawashi

If you keep your stakeholders aligned as you go along, your document won't ever have a point where you're sharing a finished document with a group of people who are learning about it for the first time. When I spoke with Zach Millman, pillar tech lead at Asana, about creating a strategy there, he told me that he used the process of *nemawashi* (*https://oreil.ly/cu6py*), one of the pillars of the Toyota Production System (*https://oreil.ly/Namw7*). It means sharing information and laying the foundations so that by the time a decision is made, there's already a consensus of opinion.[11] If there's someone who'll need to give a thumbs-up to your plan, you'll want those people to show up to any decision-making meeting already convinced that the change is the right thing to do. I've always framed this as "Don't call for a vote until you know you have the votes," but I was delighted to learn that there's a word for it.

Keavy McMinn told a similar story of a strategy she created while she was at Github. By the time she was ready to share the document with the whole company, she had complete buy-in from her boss and his boss, and she'd done a ton of behind-the-scenes prework. The decision makers already knew that the work should be staffed. Launching the document still built momentum and excitement around the work and helped a broader audience clarify the details and buy in to the decisions.

Don't forget that aligning doesn't just mean convincing people of things. It goes both ways. As you discuss your plans for a vision or strategy, those plans might change. You might realize that many people are getting hung up on some aspect of your document that wasn't really important to you, and so you end up removing it. You might compromise on some point that is a source of conflict, or give extra prominence to something that wasn't hugely important to you but that is really resonating with your audience. You might even legitimately find a better destination to aim for. All of this is OK, and is why writing a document like this takes time.

---

11 As he writes (*https://oreil.ly/zJhkc*), you want to hear the "spicy opinions" in one-on-one meetings and adjust your plan as necessary before the decision makers meet as a group.

Work on your story

A vision or strategy that not everyone knows is of little value to you. You'll know the direction is well understood if people continue to stay on course when you're not in the room to influence their decisions. But to make that happen, you'll need to get the information into everyone's brains. You can't do that if you give your organization a long document to memorize; you'll need to help them out. This is a place where the pithy one-liner or "bumper sticker" slogan I mentioned earlier can really shine. In his article "Making the Case for Cloud Only" (*https://oreil.ly/jZ9SS*), Mark Barnes writes about coming up with the slogan "Cloud only 2020" as a powerful way to make it easy for everyone at the *Financial Times* to remember their cloud strategy. Sarah Wells, speaking about the same migration (*https://oreil.ly/juTzy*), added, "It's certainly the one thing from our tech strategy that developers could quote." If your teams know, understand, and keep repeating the story of where they're going, you're much more likely to get there.

Your project is also more likely to be successful—and cost less social capital —if you can convince people that they *want* to go to the place you're describing. As you write, think about how your words will be received and be clear about what story you're trying to tell. To get back to the idea of drawing a treasure map, imagine that you've done that, and now you're in the pirate bar, rolling your treasure map out on the table and trying to make the other people at your table want to come along with you. What are you telling them?

You want a story that is *comprehensible*, *relatable*, and *comfortable*.

First, you'll want to make sure the story is comprehensible. A short, coherent story is much more compelling than a list of unconnected tasks. It's hard to make people enthusiastic about something they don't understand, and you're missing the opportunity to have them tell the story when you're not there. Even if they're brought along by your enthusiasm, if they don't really understand the plan, they can't champion it.

Make sure the story is relatable. The reason the treasure is exciting for you might not be at all exciting for other people, so the way you frame the story really matters. If your vision is that your own team will have solved its most annoying problems, live happier lives, and eat ice cream, that's pretty compelling...for people on your team. If achieving that vision will need work from other teams, you'll need more. Show how your work will make their lives better too.

Similarly, remember the Overton window and make sure the story is comfortable. A compelling story to take people on a journey from A to B will only work if they're actually at A. If they're a long distance back from there, you might have more success in convincing them of A, and then waiting until that idea is considered sensible and well-accepted before taking them on the next step.

Your story will help people as they execute on the plan too. As Mojtaba Hosseini told me, "When it gets difficult, everyone needs to know that the difficulties are expected, and that they can be overcome. Don't just tell the story of the gold at the end of the journey. When there are problems, you need to be able to emphasize that this is the part of the story where the heroes get caught in the pit...but then they get out again!"

CREATE THE FINAL DRAFT

It can feel hard to believe, when you've spent weeks or months creating a document, but not everyone will be excited to review it. Don't be offended! While people may have the best intentions, a lengthy document can stay open in a tab for a long, long time. Think about how you can make it easy to read, or share the highlights in another way. Avoid dense walls of text. Use images, bullet points, and lots of white space. If you can find a way to make your points clear and memorable, more people will grasp them.

One way is to use "personas": describe some of the people affected by your vision or strategy—developers, end users, or whoever your stakeholders are—and describe their experience before and after the work is done. Another approach is to describe a real scenario that's difficult, expensive, or even impossible for the business now and show how that will change. Be as concrete as you can. Unless you're presenting solely to engineers in the same domain, avoid jargon. Some of your readers will start tuning out after they hit a few acronyms or technical terms they don't know.

You may find it useful to have a second type of document to accompany the one that you've written. If you're going to present at an all-hands meeting or similar, you'll want a slide deck. You may want both a detailed essay-style or bullet-pointed document and also a one-page elevator pitch with the high-level ideas. If you're comfortable sharing with people outside the company, you might write an external blog post: this can be another opportunity to reach internal audiences too.

The Launch

There's a difference between a vision document that is one person's idea and a vision that is the company or organization's officially endorsed north star, with teams working to achieve it. I have seen so many documents die at this point because the authors didn't know how to make them real.

MAKE IT OFFICIAL

What's the difference between your document being *yours* and *the organization's?* Belief, mostly. That starts with endorsement from whoever is the ultimate authority for whatever scope your document needs. Usually this is whoever is at the top of the people-manager chain: your director, VPs, CTO, or other executive. If you've been using nemawashi and staying aligned with the people whose opinions matter, that person might already be on board. If so, see if they're willing to send an email, add their name to the document as an endorsement, refer to it when describing the next quarter's goals, invite you to present your plan at an appropriately sized all-hands, or make some other public gesture of accepting the plan as real. If you don't have their support, ask your sponsor to help you sell the idea.

Make sure your document looks real. Host it on an official-looking internal website. Close any remaining comments and remove any to-dos. Consider removing the ability to add comments, and leave a contact address for feedback instead. If you can include the head of the department or similar as a contact, that'll carry a lot more weight than if it has just engineers' names on top.

An officially endorsed document gives people a tool they can use for making decisions. However, there's another important part of making the document real: actually staffing the work in it. If you've proposed new projects or cross-organization work, you may need headcount—and actual humans to fill that headcount. If you'll need a budget, computing power, or other resources to make the work happen, that need should have come up in the course of agreeing on the direction, but now you'll face the reality of actually getting it. Talk with your sponsor about how to work within your regular prioritization, headcount, OKR, or budgetary processes.

Depending on your organization, you may be personally responsible for starting to execute on the strategy, or you may be handing it off to other people to make the work happen. In my experience, you'll all be more successful if you stay with it, making sure the work maintains momentum and the plan stays clear as the vision or strategy turns into actual projects.

KEEP IT FRESH

Shipping a vision or strategy doesn't mean you can stop thinking about it. The business direction or broader technical context may change and you'll need to adapt. You may also just find out that the direction you chose was wrong. It happens. Be prepared to revisit your document in a year, or earlier if you realize that it's not working. If the vision or strategy is no longer solving your business problems, don't be afraid to iterate on it. Explain what new information you have or what's changed, update it, and tell a new story.

Case Study: SockMatcher

Time to go back to our friends at SockMatcher. When you read the scenario at the start of the chapter, maybe you had ideas for what they should do. In some ways, the technical problems are easy to solve. But making a change that affects many people is anything but.

Let's imagine you're a staff engineer working at SockMatcher. Here's the story of your approach, writing, and launch.

APPROACH

Your previous project has just wrapped up and you're looking for something impactful to do next, ideally something that's a bit of a stretch. Creating a plan for the most contested core architecture in the company certainly fits the bill. It also feels like an important problem, one that can have a huge impact on the business.

Your manager is wary. Many others have tried to tackle this architecture before—this may be an uncrossable desert! You suggest that you take a couple of weeks to understand why previous attempts failed. If you don't have a compelling reason to believe that your journey can be different, you won't do it.

Why didn't previous attempts work?

You start out by chatting with two staff engineers who have taken a run at rearchitecting the monolith in the past.

The first, Pierre, spent three months creating a detailed technical design for the monolith and surrounding architecture. Other teams weren't impressed: they disagreed with some trade-offs Pierre had made, the direction didn't match their own plans for their components, and they didn't like having a solution handed to them to implement. Unable to rally enthusiasm, Pierre decided the project couldn't be solved (at least, not with the current set of engineers). He's still pretty grumpy about it.

The other engineer, Geneva, set out to build a coalition before attempting a rearchitecture. She set up a working group and there was a ton of interest. The various participants were initially eager to work together, but the working group got bogged down in debate and couldn't agree on a path. The hours of meetings became a time sink and people stopped going, including Geneva.

As you talk with these and other engineers who have opinions about "solving" the monolith, you notice two patterns. The first is that most people have a specific solution in mind: "The problem is that we don't have microservices" or "We just need to shard the data stores." Everyone wants their solution to "win," so consensus is impossible. The second pattern is that everyone is focused only on the technical problems. There are lots of technically sound ideas, but no plans for how to get the organization to buy in to a path forward.

You think you can be more successful if you tackle organizational alignment as the crux of the problem. You resolve to get an executive sponsor and make sure that any directions you propose are not just good technical solutions but are viable within *this* organization. You'll be pragmatic and low-ego, helping existing ideas succeed rather than trying to have your own direction prevail.

Sponsorship

You've got some social capital and credibility (see Chapter 4) in the bank after your last project, but you know that's not enough to convince all of the many teams that care about the monolith. You also know that you're going to have to make some decisions and you won't be able to make *everyone* happy. If you need complete consensus, you're not going to succeed. Also, any plan you make is likely to create engineering projects. If you're not going to be able to staff the work, you'd rather find that out early, before you waste your time. You need an executive sponsor.

You start with Jody, the director whose teams operate the monolith: she has a vested interest in making it easier to maintain. But she's seen her people get pulled into the two previous attempts to change the architecture, and she wants to defend their time. They have their own projects, and she doesn't want them to be distracted by yet another new initiative. While she's in favor of rearchitecture, it's in a "Next year, we hope, maybe" sort of way. She's not interested in committing anyone to this work.

Your next stop is Jesse, the director who's taking on the food storage container launch. With this high-profile project coming online, Jesse might have easier access to staffing and, if you can align his success metrics with your own, he's likely to support the work. When you talk to Jesse, you describe a future where product teams can work autonomously, product engineers are happier, and new features are delivered quickly. Jesse isn't sure. That's a nice future, but food storage needs to launch this year; they can't wait for a massive rearchitecture. You agree: any solution must let the food storage folks launch with minimal friction. Jesse is convinced. He agrees to sponsor and support your work.

Other engineers

You look for coauthors, a few colleagues who can bring different perspectives and knowledge to the work. You also want to build allies across your organization and engage anyone who will be skeptical of or pull against your plan.

You start with Pierre, the staff engineer who proposed the detailed previous solution. He's still feeling a bit raw from putting his heart into making a thorough solution and meeting with complete apathy. He says some defeated (and kind of mean) things about the company leadership and makes it clear that he thinks your work is a waste of time. You ask if you can use his previous plan as an input to some of the work you're doing, crediting him for any parts of his work you end up using—though you set expectations that you'll scope your project differently. The idea of his work getting used makes him a little more willing to help. He still won't join your group, but he agrees to be interviewed and to review your plans later.

You have higher hopes as you invite Geneva to join efforts. She's in. You're surprised when she tells you that her previous working group still exists, sort of: three senior engineers meet and talk about architecture every week. They've given the monolith problem a ton of thought, and you know they'll be able to see nuances that wouldn't be immediately obvious to you. You invite them to team up, asking them to commit two days a week for at least two months. One engineer, Fran, agrees; the other two want to advise but can't commit a big block of time. You agree to come back to the working group meeting every few weeks with updates.

You check in with some other potential allies:

- The team lead on the food storage container project is inclined to see problems as easier to solve than they actually are, especially when it comes to work assigned to the monolith maintenance team. "Why don't they just build isolated modules within the monolith?" she asks. But if there's a plan, she's on board.

- The databases lead is wary that your project will drop unexpected work on his team. (It's justifiable: there's history.) You promise to keep him in the loop and let him review plans early on.

- The staff engineer who wrote the original sock matching code is very tenured and *very* influential. If she's convinced, a lot of other people will be too. She has some ideas and wants to be an early reviewer, too.

Scope

Your core group—you, Geneva, and Fran—talk about what you want to do and what might be successful. Fran is eager to create a technical vision for how all of your architecture evolves, but it's not the right choice for your situation: you don't have that scope of influence and neither does your sponsor. Also, a project of that scope couldn't be ready in time for the food storage launch.

What about a vision for the core monolith architecture? That would clarify where you're all going, but teams would still be divided on how to get there. You decide you need a broad architectural plan for the monolith and a strategy for how to get there that explicitly includes the food storage launch. You'll aim to describe one year's work and stay at a high level. You commit to making occasional lower-level technical decisions, but only when the decision doesn't have an owner: you'll leave most of the implementation details to the teams that will do the work. And your strategy has to be official: it can't be just *a* plan; it has to be the organization's chosen plan or you won't consider the project a success.

Once you're all on the same page, you write down what you're going to do: "Create a high-level one-year technical strategy for enabling the food storage launch while evolving our core monolith architecture." It's a little vague, but it's a start!

You talk more about the scope and the problems, collecting links to the previous efforts and the working group's notes, and getting yourselves (and your shared vocabulary) aligned. Your document is rough and not something you'd share outside the core group, but it keeps your ideas in one place and lets you all add extra thoughts as they come to you.

Once you have an elevator pitch for what you're aiming to do, you check in with Jesse, your sponsor. He's on board with the scope and the kind of document you're creating. His suggestion for making your plan official is to add an organizational OKR for it, with your name as the directly responsible individual (DRI). That's a little intimidating, but it will certainly give you the official endorsement you were hoping for. He offers to make sure Jody and the other directors are comfortable with it, and to get the OKR added.

You create a discussion channel for the effort, announce it in other channels that are likely to have interested parties, and share notes about your scope and what prior art you're drawing from. You highlight that you want to talk with people who have opinions, and are still welcoming collaborators who have at least two days a week to spend on it. There are a *lot* of the former—and none of the latter. You start listing people to talk with and set out to write your strategy.

THE WRITING

Before you jump into solutions, you want to be *really* clear about what problems you're solving. Your initial scope of "Create a high-level one-year technical strategy for enabling the food storage launch while evolving our core monolith architecture" needs more clarity, but you don't want to jump to solutioning either. You need to diagnose the situation and describe exactly what's going on.

Diagnosis

There are so many facts that you could consider:

- There's an immediate product need to support food storage containers, and there are indications that product lines will expand more in future.
- The team running the monolith is getting paged too much: that's not sustainable and needs to change.
- Your matching algorithm is a little slow and could have a higher hit rate.
- Your systems are not currently handling spikes in traffic.
- Users are unhappy with your availability.
- The login system is old and has a lot of technical debt.
- Deploying new code is slow and frustrating.
- External users depend on APIs you wish you could change.

- Adding new products to match requires major logic changes in several core components.

- Many developers just don't like working in the monolith.

And that's not even the entire list! You set out to select the facts that matter most and tell a much simpler story. Your group spends some time brainstorming about what's important, what's not working, and what's great as it is. You imagine your future selves looking at the same codebase with twice as many engineers and another five products. If those products were implemented as edge cases too, navigating the business logic for each feature would become horrific, and changing anything in that scenario would be complicated and fraught. Builds would be slower and deploys would fail more often. More celebrity users would mean more outages. If you take no action, this is the future. You'd better take some action!

Although you've got lots of ideas after your brainstorming session, you don't want to get too committed to them just yet. Instead, you go chat with some members of your product teams, as well as engineering leaders and practitioners, aiming to understand what's important to them. You learn some new information:

- While you'd speculated that traffic spikes from celebrity endorsements might be causing the decrease in availability, outages caused by overloads are actually uncommon and brief: just a few minutes of downtime per month. While you should certainly improve here, these outages are not the most important contributor to your poor availability.[12] It turns out that the real damage is being caused by unremarkable code bugs, and by the fact that it takes three hours to deploy a fix. Previous attempts to improve reliability have centered on adding more testing to your release path—but this actually slowed deploy times and lengthened outages.

- Many engineering teams complain about working in the monolith, but the thing they actually hate is releasing code. A huge percentage of changes have unexpected behavior. It's hard to be sure that your change isn't breaking someone else, and getting a fix out takes half a day. Teams are twitchy from responding to outages.

---

12 If dropping some of this celebrity traffic was considered to be bad PR or a missed opportunity, you might prioritize it anyway. Context is everything.

- The billing and personalization subsystems are by far the most contested parts of the codebase. Most major feature changes come with a corresponding change in one or both of those pieces of functionality, and their logic is so complex that it's easy to have unexpected side effects while making even simple changes.

You learn a lot more, too: everyone you talk to has a different topic they want to tell you about. But the interviews build up a pattern and let you see what's going on. You choose the most important subset of the facts and tell a much simpler story.

Here's your diagnosis:

Every new feature change needs complex logic changes in a set of shared components. Modifying these components is slow and difficult. Unexpected interactions mean that teams are constantly disrupting each other's work and causing long outages. Every new type of matching item will increase the number of points of coupling inside these shared components and make the problem worse. Our systems need to be able to handle more matching components and more teams adding them without development grinding to a halt.

Choosing your focus can be one of the most painful parts of writing a strategy. In this case, the unversioned APIs are still a problem, the unpleasant login code will become a problem eventually, and improving the match functionality is a real opportunity that you're not going to try for right now. Those problems and opportunities are real, but you're making a hard decision and ignoring them for now.

With the diagnosis, you check in with your sponsor, Jesse, and make sure he agrees that you're focusing in the right area. You show him the list of challenges you're *not* focusing on too, to make it clear that you see them but don't think they should come first. Jesse agrees.

Guiding policy

Now that you've got a clear sense of what's happening, you can decide what to do about it.

There's a proposed guiding policy already on the table: some of your colleagues are pushing to completely break down the monolith. They say that the rearchitecture would make it easier to add new products, as well as reduce the number of unintended breakages and the time to get a code fix built and deployed when something's broken. But teams would still need to make risky changes to shared components. It would also mean that all of the teams would begin running their own services, putting some of them on call for the first time in their lives. Finally, a change like that would also take at least three years, with no solution for shipping products in the meantime. So, while "let's run microservices instead" might be the perfect solution for a company with different constraints, it doesn't acknowledge the current situation.

Instead, you look at the places where a small amount of work can have an outsize impact. An obvious point of leverage is the two key shared components where integration slows teams: billing and personalization. If those two were easy to add to, instead of having hardcoded logic for each product, other teams could safely add new kinds of matchable items. There would be fewer outages, freeing up teams to spend their time on feature work and improving the system further. It's a virtuous cycle.

Here's your guiding policy:

The billing and personalization systems should be easy and safe to integrate with.

You write up some notes about the guiding policies you *didn't* choose, too. You describe the reasons you considered microservices, and the advantages that path would bring, but explain why they wouldn't solve your problems.

Actions

You set out to outline the actions your group will need to take to navigate the challenges and carry out your guiding policy.

It's important that your actions are realistic, so you check in with the billing and personalization teams and leadership and confirm that they're on board with your direction. The billing team already had some backlog work around offering a menu of billing functionality that other teams could choose by setting configuration options. The personalization folks had toyed with the idea of a plug-in architecture with stable core functionality and isolated logic for each type of matchable item. Teams adding new items would only have to modify their own plug-in component. Both of these changes would make the shared components more modular and enable self-service access to them, and the teams would be happy for a reason to spend time on them.

There's a bootstrapping problem, though: these are big, risky changes. Refactoring these core components is likely to cause many outages, so the teams have not prioritized the work. You suggest adding the ability to release changes behind feature flags (*https://oreil.ly/ErwHv*), so a regression can be a quick switch back, rather than another deploy. Safer deploys would reduce the cost and risk of the work.

The isolation work won't happen overnight, and the food storage team will need integrations with both these components in the meantime. The personalization and billing teams are willing to treat food storage as a pilot customer and optimize for making them successful, including writing the first version of their integrations in their existing systems and migrating them to the self-service model when it's available. But those teams can't take on both isolation and integration work at once with their current staffing. Jesse agrees to donate some of the headcount that had been allocated to the food storage project and let both those teams grow.

Here are your actions:

- Add a feature flagging system that allows staged rollouts and quick rollbacks.

- Add two engineers to the payment team and one to the personalization team.

- Modify the billing and personalization subsystems to allow easy, safe, self-service additions of new matchable items.

- Have the billing and personalization teams onboard the food storage product into their systems, then migrate them to be pilot customers of the new self-service approach.

These actions are high-level and the teams involved have autonomy to design solutions and make a lot of decisions. But this approach gives them a direction and some concrete next steps. There are many, many other suggestions for actions: everyone you talk to has a laundry list of the work they think should happen to make the monolith healthier. But your guiding policy lets you focus your efforts and keep your list short.

You nominate Fran as the primary author to write up the plan, with you and Geneva making many suggested edits. The document is honest about the trade-offs of your plan as well as the alternatives you considered and why you didn't choose them.

After you've aligned with your sponsor (he suggests some wording changes but is overall enthusiastic), you road test your plan by sharing the first draft with a few of your allies. Some leave comments. You interview others in person and shake out some concerns.

Your story becomes a little tighter every time you tell it. You share the document with progressively broader groups and start to present about it at some meetings.

THE LAUNCH

Let's be honest: your plan is not universally beloved. Some of your colleagues are underwhelmed (and, perhaps, a little angry): your document isn't more "visionary" than any of the ideas they had; it's a little boring! Some insist that this problem didn't need a strategy; you just needed to "just decide" what to do, and this path was the obvious decision.

Other vocal factions disagree that your path is obvious: they think it's just *wrong*. One group is still arguing for moving everything to microservices. Another wants more focus on handling load spikes. And while the food storage container team can build some of their functionality as a microservice, they'll still be developing heavily inside the monolith—some of them are unhappy about that. However, since you took the time to document the known disadvantages and the alternatives you considered, none of this is news. The grumbling doesn't change the plan.

Happily, the positive voices are louder. Most people are energized by having a single, official, agreed-upon direction. You have particularly strong buy-in from allies who were along on the journey from the start. Your engineering-wide OKR has raised your visibility, and your sponsor and his peers are aligned with the work and willing to staff it. Your plan will take some load off the monolith maintenance team without needing a huge commitment from them, so Jody unexpectedly offers some of their help: they'll provide the feature flag system.

You stay with the project for the next year, working directly on the billing modularization project and acting as an adviser for the personalization work. Immediately after the food storage team celebrates a successful launch, the product team announces a new effort to match missing board game pieces. Your work means that the board games team will be able to use the new isolated functionality and just start building. With stable systems, the monolith maintenance team is no longer reacting constantly: they've begun to work on improving the developer experience and making the systems more resilient to load spikes.

By focusing your efforts, you removed an impediment for growth, and a barrier that was slowing everyone down. And you're ready for another project.

Onward to Part II of this book, Execution.

To Recap

- A technical vision describes a future state. A technical strategy describes a plan of action.

- A document like this is usually a group effort. Although the core group creating it will usually be small, you'll also want information, opinions, and general goodwill from a wider group.

- Have a plan up front for how to make the document become real. That usually means having an executive as a sponsor.

- Be deliberate about agreeing on a document type and a scope for the work.

- Writing the document will involve many iterations of talking to other people, refining your ideas, making decisions, writing, and realigning. It will take time.

- Your vision or strategy is only as good as the story you can tell about it.

PART II

Execution

Finite Time

When I was new to the industry, I used to wonder why my senior colleagues seemed so wary of committing to things. I would demonstrate that something was a problem and, inexplicably, my boss and the senior engineers around me would not drop everything and go solve that problem. Why didn't they care?

Now that I'm the senior colleague, I get it. As a senior person, you can see a host of problems: architectures that won't scale, processes that waste everyone's time, missed opportunities. When someone points out one more, you add it to the list.

The good news is that you're unlikely to run out of work. The bad news, though? You can't do it all. While you might like to jump on every problem, you quickly learn that doing that is just not sustainable.[1] You have to make peace with walking past things that are broken or suboptimal (or just really annoying) and taking no action.

Doing All the Things

In this chapter, we'll talk about choosing what to do. As a staff engineer, you'll have choices to make every day: whether you should join an incident response call, how you'll respond to a request for mentorship, whether to take on a particular side project. Not everything can be your problem. So what should be?

We've already looked at opportunity cost and only taking on work that's important for the company, but in this chapter I want to add an extra layer: what's important for *you*. Choosing projects that support your growth, reputation, and happiness can feel a little selfish in the short term, but your needs are important, and you're the person with the most incentive to watch out for them.

---

1 OK, not that quickly. It's really difficult. But you eventually learn.

Your time is the most obvious finite resource: there are exactly 168 hours in every week, and that's what you get for the rest of your life. We'll start there, then look at five other resources that you need to manage: energy, quality of life, credibility, social capital, and skills. These five will carry different weights depending on the stage of your career, your recent successes, and what else is going on in your life. We'll look at how new projects can affect them and how you can deliberately choose a project engagement "shape" that makes sense for you.

To conclude the chapter, I'll work through some examples of the kinds of work that you might consider starting, weighing up their value to the company and to you. We'll look at ways you can amplify the positives, decrease the negatives, or just say no and decide not to do them.

There's infinite work. There's exactly one of you. Let's start by looking at your week.

Time

Everything you commit to has an opportunity cost. By choosing to do one thing, you're implicitly choosing not to do another. If you use five minutes between meetings to fix a dead link in your documentation, you're choosing not to reply to an email or get a glass of water just then. If you agree to spend the next two years on *this* big impactful project, you aren't available for *that* one.[2] No matter what the scale, the decision has a cost.

FINITE TIME

I'm inclined to be optimistic about my time. Too optimistic. There's always important, interesting work available—and my default is to say, "I'll do that. I'll fit it in somehow." I've had to work to be aware of this tendency and to remember that I have finite time.

It's an odd default that work calendars tend to only show meetings. If you're trying to do focused work, your planned schedule only shows the *interruptions* to that work, not the work itself. Executive coach Fabianna Tassini of Confidantist gave me the best advice for managing my workload: *put nonmeetings in the calendar too.* And I don't just mean big blocks of "make time" for other people to

---

2 In this chapter I'm going to use the word *project* very loosely, to mean any initiative or standalone task, from answering a colleague's question to a huge cross-organizational effort that's intended to last for years.

apologetically schedule over; I mean specific, deliberate items.[3] It's a powerful way of visualizing what's going to get done and when. See Figure 4-1 for an example.

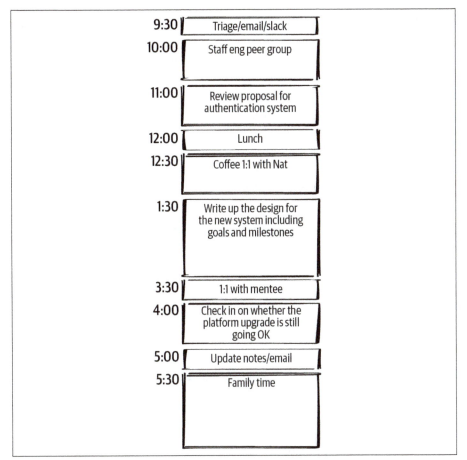

9:30 Triage/email/slack

10:00 Staff eng peer group

11:00 Review proposal for authentication system

12:00 Lunch

12:30 Coffee 1:1 with Nat

1:30 Write up the design for the new system including goals and milestones

3:30 1:1 with mentee

4:00 Check in on whether the platform upgrade is still going OK

5:00 Update notes/email

5:30 Family time

Figure 4-1. A day in the life. The calendar shows both meetings and focus work.

Having my work in my calendar means I can see whether I have time for a side quest. I love side quests, so it's helpful for me to be confronted with the

3 As I write this, my calendar includes making slides for an upcoming all-hands meeting, reading a design that we're going to discuss next Monday, catching up on a long and nuanced Slack thread, and texting the guy who's supposed to come stop my roof from leaking. In the past I might have squeezed these tasks in around the meetings, but all four of them are more important than most of the meetings I have this week.

reality of how I've planned my time and to ask, "If I start this work, what am I not doing instead?" When someone asks me to review their 20-page document, I can honestly tell them when (or whether) I'll have time. If it's more important than something I've already scheduled, I can move that other thing and make space. If I find that I've rescheduled the same piece of work multiple times, I'm getting an indication of how important it is. Am I avoiding this thing? Might it actually not...matter? Maybe I can stop trying to do it.

Calendars are great for days or weeks, but if we're looking longer term, we need a bigger picture. Figure 4-2 shows a sketch (I'll call it a *time graph*) of a month or a quarter before any projects go into it. I've blocked off a fraction of the time for the kind of activity that is just the background noise of working in a corporate environment: all-hands meetings, performance reviews, and so on. Depending on the enterprise, this will be a bigger or smaller chunk, but most likely some amount of your time is already spoken for.

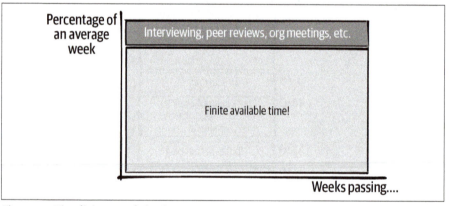

Figure 4-2. Visualizing your finite time.

The rest of that time graph shows the hours you have available for project work, long-term planning, reviewing code and documents, building relationships, keeping up to date with what's happening in the company and the industry, mentoring, coaching, nudging projects in the right direction, and learning new technologies you'll be expected to have an opinion on. And having lunch. Every time you take on something new, you're adding a block to that graph. Maybe it's a huge block. Maybe it's a tiny dot. Either way, it's taking up space.

HOW BUSY DO YOU LIKE TO BE?

Leadership work can be unpredictable. A crisis, outage, or launch can cause a load spike. If a project needs more help than you predicted, you might find yourself oversubscribed. So, when you're filling your schedule, think about how volatile your incoming workload might be.

If you allocate 100% of your time and something unexpected happens, your choices are to drop something or run beyond capacity. If a lot of your tasks aren't time-sensitive, dropping things might be easy. But if you fill your schedule with only important things, then when you hit your limit, by definition you're dropping something important. If you decide *not* to drop anything, then work life will inevitably spill into other areas of your life, causing stress and exhaustion.

Know how many hours you want to work on an average week, how many you're comfortable spiking to, and at what point you'll stop being able to handle the load and fall over. I know people who run like the "A" person shown in Figure 4-3 and are completely unruffled when a crisis or an opportunity means they want to put in a few extra hours. I know others who work like person C, always right at their maximum capacity and stressed out all of the time. Try to leave at least a little buffer space if you can.

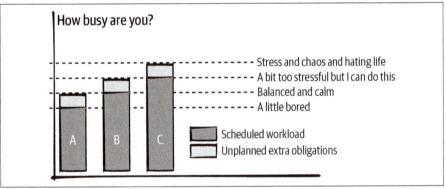

Figure 4-3. *How full is your work schedule? If something extra happens, can you handle it without melting down?*

PROJECTQUEUE.POP()?

You've got a *locator map* to give you perspective about what's important and a *treasure map* to remember what you're trying to achieve. Using those, you should be able to evaluate any project's importance. In theory, then, should you be able to sort the available projects, like in Figure 4-4, and have each engineer continually take the next item from the top of a priority queue?

Possible work to do in strict priority order

- Lead project to replace storage layer
- Create frontend strategy
- Mentor new data engineer
- Design API for customer platform
- Introduce Mia to Adam
- Data privacy training
- Implement new feature

Figure 4-4. A list of work sorted by priority. If a five-minute introduction is backed up behind all of those projects, those people aren't going to meet for a while.

I think we can all see that that would be a little silly. This list mixes big and small work, and some engineers will be a better fit for some projects. There will always be a balance between choosing the strictly most important next thing and choosing the work that's right for you.

So this next section will be about how to evaluate how well a project fits with your needs. Note that this section will be a little selfish! You shouldn't use these rubrics in isolation: if you're looking at something you might want to do, I'll assume you've already thought about whether the project is important, useful, timely, achievable work that fits with your organization's needs and culture.

But you are not an interchangeable cog in this organizational machine, and taking care of your needs is compatible with being a team player. In fact, if you look for ways that your projects can keep you healthy and happy and building your skills, you're going to do better work—and it will be easier for the company to retain you. Everyone wins. So let's talk about your needs.

Resource Constraints

If you've ever played the person-simulation video game *The Sims*, you'll remember that each little simulated person (a Sim!) has a dashboard (see Figure 4-5) showing their levels of comfort, energy, social life, etc. The needs bars increase or decrease in various situations, and a big part of playing the game is keeping your Sim in good shape, giving them activities to increase their level of "fun," giving them enough sleep to maintain "energy," and so on. If one of the needs is really in the red, your Sim gets into a terrible mood, and some activities just aren't available to them, even activities they'd benefit from.

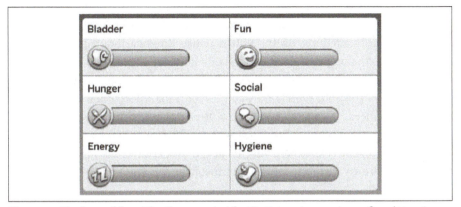

Figure 4-5. Needs panel from The Sims 4, copyright EA Games (source: image from https:// www.ign.com).

YOUR DASHBOARD

OK, you're probably not simulated. You're probably a real human.[4] But still, can you imagine a little dashboard for yourself (Figure 4-6), showing your current levels of various needs? Imagine it includes five resources: energy, credibility, quality of life, skills, and social capital.[5]

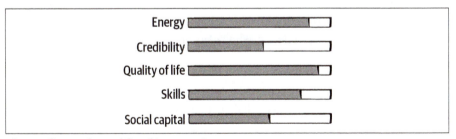

Figure 4-6. A needs panel for taking on projects.

Let's examine each of these needs and what increases and decreases them.

---

4 If you are a sophisticated AI, of course you're still welcome here. Thank you for choosing this book.

5 These categories are far from comprehensive, but they're a useful model for thinking about projects. What else would you put on your dashboard?

Energy

In theory, if you have a free hour, you can choose to spend it in any way you want. In practice, it will depend on how much energy you have. It's easy to run out of what in my family we call *smartbrain*: the energy to stay focused on a piece of work and do something useful with it. Once the smartbrain is gone, it's a struggle to do anything useful. At the end of a long day of meetings, I sometimes have a free hour that I could use for reading documents, but my brain is mush: I can parse the words, but there's no way I'll retain the information or notice any unspoken nuances. I'll have to spend five times as much willpower to stay on that tab and not drift into reading Twitter. And if I try to write, I won't be able to put thoughts together: even replying to an email becomes an insurmountable challenge. There's a barrier to entry: you must have *this much* energy to be able to start this task.

Different people are energized or exhausted by different things. I'm absolutely wiped by one-on-one meetings, but I have friends who thrive on them. I can code or write with no obvious internal limit once I've gotten started, but after about an hour of reading documents or debugging, my brain starts shutting down and my eyes glaze over. Understand what kinds of work are expensive for you, and what kinds will leave you with some smartbrain at the end of the day.

Your energy will be affected by factors outside work. If you've got a baby that's not sleeping, you're going to start the day with less energy in the tank. If you're moving to a new house, dealing with illness, or living through ongoing stressful situations, those will take their toll on your energy too.[6] Very few of us can compartmentalize our energy: when you go to work, you're still the same you.

Quality of life

In tech, most of us are in the very privileged position of being able to do work we enjoy and choose. The work can be hard, but it's intellectually stimulating, it tends to be well paid, and it's usually not dangerous. We're very lucky. But it's possible for this to be true and for the work you're doing to still make you deeply unhappy. Certainly not all of your quality of life will or should come from your work; you may even stick with work you dislike because it's a step toward something you want and you're optimizing for future happiness. (See Chapter 9 for

---

6 At the time of writing, we're entering year three of Covid and I don't know anyone who's at maximum energy.

more about this.) But we spend a lot of our lives at work, and it's reasonable to want to feel good about it.

If you enjoy the kind of work you're doing and the people you're working with, that will be a boost to your quality of life every day. If you're bored or working with people who treat you badly, your job might chip away at your happiness instead. Your quality of life will also be affected by other resources: if a project eats up your energy, you might not be able to do other things you enjoy. If it boosts your profile and makes people admire you, the recognition can feel good. You'll be affected by whether you believe in the journey you're all taking: the most interesting technology and enjoyable coworkers might not compensate for feeling that you're doing harm in the world. And, of course, money can have a massive impact on your quality of life, and that of your family or dependents too. As one person I spoke with said, "Working for a megacorp and getting the biggest bucks might not make me personally happy, but it would let me pay for my mother's elder care, and that might be something I weigh above personal happiness."

Credibility

As a staff engineer, it can be easy to drift up to higher-altitude problems and feel less "in the trenches" with the technology. This is not inherently a bad thing: there's work available at all altitudes. But if you move entirely away from low-level technical problems, other engineers might distrust your technical judgment because they think of you as too disconnected. In some cases they might be right! Your understanding of what's possible and what's good practice might get out of date.

You can build credibility by solving hard problems, being visibly competent (see Chapter 7), and consistently showing good technical judgment. When I asked on Twitter (*https://oreil.ly/BzVVW*) what causes a loss of credibility, one of the big themes was absolutism: if you're a fan of some technology and advocate for it in every single situation, people will stop believing you know what you're talking about.

Credibility extends to your skills as a leader, too. If you're polite (even to annoying people), communicate clearly, and stay calm in stressful situations, other people will trust you to be the adult in the room. If you are rude or highly dramatic, send emails that are unreadable walls of jargon, or make all-hands meetings wait while you ask a rambling question that only applies to you, it will have the opposite effect.[7] You will build credibility as a professional every time you take on a chaotic situation and make it easier for everyone else to understand. You'll lose credibility when you're seen as contributing to the chaos, or when a project goes badly and you don't do a good job of navigating the failure.

Credibility is another resource where there's often a bar to entry. You won't be offered a difficult project or opportunity unless someone believes you can succeed at it. And any change you propose will be more welcome if other people believe you know what you're talking about.[8] As Carla Geisser says in her article "Impact for the Impatient" (*https://oreil.ly/MwJ5B*), "the Giant Maybe Unsolvable Problem will be easier after you've shown you can get things done."

At staff+ levels, you're often trying to find a balance between seeing the big picture and accepting pragmatic local solutions. If other engineers see staff engineers as working in an ivory tower and advocating for work that doesn't feel valuable, it's even more important to be aware of your "credibility score" and establish that you know what you're doing. But it's a fine line: if you ignore the big picture and business needs, you'll lose credibility with your leadership.

Social capital

While *credibility* is whether others think you're capable of doing whatever you're trying to do, *social capital* reflects whether they want to help you do it.[9] The term comes from sociology, where it refers to the connections between people.[10] In business terms, though, we usually look at it like this: if someone asks you to do something inconvenient to help them, do you say yes? It probably depends on how much they're in your good books. Do they help you out a lot, or are they continuously asking you for favors and giving nothing in return? Did you end up regretting the last time you helped them? Whether we talk about it or not, everyone has a bank account of capital with each of the people they know. If someone has built up credit with you, you're more likely to do them a favor or give them the benefit of the doubt. Social capital is a mix of trust, friendship, and that

---

7 There's a good chance these three people are some kind of cosmic construct present in every organization simultaneously.

8 Note that we make assumptions about other people's abilities, and implicit bias plays a part when we're deciding how credible someone is. If you get extra credibility for free because of your demographic, think about whether you can use that freebie to boost other people who don't.

9 If you're an RPG player, you could think of credibility as WIS and social capital as CHA.

10 Read Jane Jacobs (*https://oreil.ly/aGbzy*) and Pierre Bourdieu (*https://oreil.ly/qOa1K*) if you'd like to know more.

feeling of owing someone a favor, or of believing they'll remember that they owe you one.

Social capital builds up over time, and you'll need more of it with some people than others. In general, you'll want to stay on good terms with the people in your reporting chain and build a track record of helping them achieve their goals. If there's a business-critical problem and you refuse to help, or you take on an important project and don't complete it, you'll burn goodwill. And if you always ask for favors but never repay them, you'll start to find it difficult to get people to help you.

As you spend time with people, have good conversations, work together with them, help them out, make social connections, and support each other, goodwill and social capital will build on both sides. Completing projects will build capital too. If you delivered the project that made the company successful last year or unpicked the impossible architectural knot that was slowing everyone down, you'll have a lot more leeway the next time you're asking for something. As Alexandre Dumas said, nothing succeeds like success.[11]

Once you have social capital banked, spend it deliberately. When your star is high, you can often get away with chartering an initiative that other people don't really believe in, just on the strength of their faith in you (or their desire to keep you happy). Invest your social capital wisely. If you waste your "one unreasonable request" token, you won't get it back.

Skills

The skills resource behaves a little differently from the others, because it's always slowly decreasing. That doesn't mean you're forgetting what you know (though you will lose fluency with technologies or techniques that you don't use for a long time); it means that any technical skill set slowly becomes less relevant and eventually gets out of date. Our industry moves fast. If you take on projects that teach you nothing (or at least nothing that's relevant to the projects or roles you want), you won't keep up with the rate of decrease.

---

11 He actually said "Rien ne réussit comme le succès," but it amounts to the same thing.

As you work through your career, you'll increase your skills bar in three main ways.

The first is by deliberately setting out to learn something: you take a class, buy a book, or hack on a toy project. While this kind of structured learning can often happen at work, you may struggle to find time for it, and find it spilling into your free time.

The second way is by working closely with someone who is really skilled. Being the least skilled person on a team of superstars will teach you more than being the best person on an otherwise mediocre team. When you work with great people, it's almost impossible to avoid becoming greater yourself.

The third way—and the most common, I think—is learning by doing. You get better at what you spend time on. If there's a skill you want to hone, the easiest way to practice it will be to take on projects that need that skill.

There's often a bar to entry for projects and roles: you won't have a chance to take on the project unless you have the appropriate skills. So, depending on what you work on, you might be increasing your relevant skills every day or watching them slowly erode.

E + 2S + ...?

When you're considering taking on a project, there's more to the equation than whether the work is important for your company. You need to pay attention to your dashboard of needs too. When you look at how you fill your time graph, pay attention to what any new project, task, or initiative will do for each of your resources: your energy, quality of life, credibility, social capital, and skills. (See Figure 4-7 for an example.)

Unfortunately, I can't give you a formula to plug numbers into: at different times you'll optimize for different resources, depending on your current levels of each one and what you're hoping to do. Some projects will be out of reach unless you have a minimum level of some of these resources. Some life goals will too.

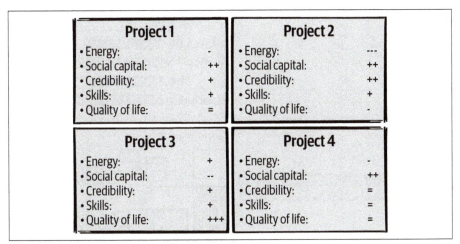

Project 1		Project 2	
• Energy:	-	• Energy:	---
• Social capital:	++	• Social capital:	++
• Credibility:	+	• Credibility:	++
• Skills:	+	• Skills:	+
• Quality of life:	=	• Quality of life:	-
Project 3		**Project 4**	
• Energy:	+	• Energy:	-
• Social capital:	--	• Social capital:	++
• Credibility:	+	• Credibility:	=
• Skills:	+	• Skills:	=
• Quality of life:	+++	• Quality of life:	=

Figure 4-7. Comparing projects based on their effect on each resource.

Not every project and task you take on has to be ideal, or even good for you. Sometimes you'll choose a project that's terrible on one axis because it's a good choice for some other reason. Maybe you take a stint of working longer hours than you would prefer because helping your team get through a crisis is important to you. Or maybe you work on the project that's most important to your boss, with the understanding that you're going to take time afterward to deep dive on an idea you've had. Sometimes a project will even be objectively terrible for you and you'll have some reason to do it anyway. But over the long run, you need to make sure your needs are being met.

BIN PACKING

Since projects can take a lot of different shapes and sizes, filling your time efficiently isn't easy. Figure 4-8 shows how complicated the decision can be. Adding any of these blocks might mean that another, more important block can't fit. And of course you have the added complexity of keeping all of the resources in good shape.

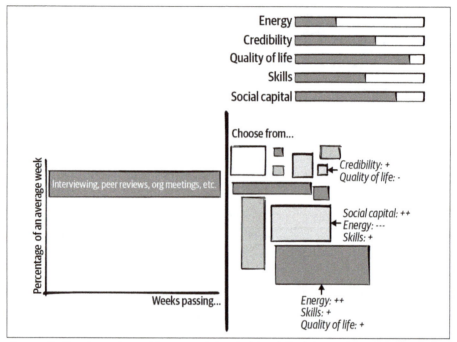

Figure 4-8. Choosing projects with your needs in mind is a difficult bin-packing problem.

It's a multidimensional bin-packing problem (*https://oreil.ly/FOpGv*). This is a famously difficult thing to optimize, so don't *entirely* overthink it, especially when the task in question will only take an hour or two.[12] Any given project will never be the be-all and end-all of your dashboard scores. But the bigger the project, the longer you should spend thinking about whether it's a good fit.

Choosing Projects

New projects and tasks will become available every day. If you always say yes, you're unlikely to get anything completely finished. If you never say yes, you'll miss opportunities that would have been good for you, and you'll look like a slacker. You can fit a lot of small tasks into a day, and you usually have more than one project on your plate at a time. But how many is too many? How do you decide?

---

12 I'm aware of the irony of writing this after clearly overthinking it for a chapter.

EVALUATING A PROJECT

Engaging with a project can take lots of shapes. But it's easier to plan and evaluate your work if you're clear about what you're signing on for.

Let's look at some places projects come from, and some of their characteristics.

You're invited to join

A project needs a lead or another contributor, and someone asks you. The project is already underway, so it may already have momentum, and you probably won't need to convince your organization to care about it. And it feels good to be sought out! But the project might not be what *you* think is most important, or the kind of work you'd enjoy doing right now, and the fact that someone thinks you'd be a good fit for it *might* mean that it's very similar to work you've done before, and not an opportunity for growth.

You ask to join

If there's a project you think would be a great growth opportunity, teach you skills you want, let you work with people you enjoy, or just be really fun, you can go ask to join it. For a lot of folks, asking is so obvious that this recommendation doesn't feel worth saying. Some of us, though, prefer to hint at our availability and wait to be invited. If you've ever done that, this public service announcement is for you: even if it's obvious that you're the best fit, don't wait for someone else to notice. Go ask for the project you want.[13] Everyone will be happier and you'll increase your chances of actually getting it.

You have an idea

Sometimes opportunities for new projects jump out at you. Working on your own initiative likely means you'll need to find the words and narrative to convince other people to care, especially if you want them to work with you. It can be frustrating to find yourself mired in navigating the organizational structures to

---

13 OK, I talked in Chapter 2 about organizations that have crystallized, where asking for a good project before your turn will be frowned upon. If you're in one of those orgs, make up your own mind about whether this is terrible advice. I'm also conscious here that this dynamic will be influenced by culture, gender roles, and other factors. Alex Eichler's *Atlantic* piece about ask versus guess cultures (*https://oreil.ly/JVH8d*) is a good read. All that said, unless it *really* feels impossible, register your interest in any project you want. Both you and the person looking for a lead will lose out if they don't realize that you're interested in it.

create a new project, get headcount, justify the business case, and show results, when you just want to get into designing or coding. But once it's underway, you get to lead a project the way you want it to be led and maybe have a huge impact.

The fire alarm goes off

There's a critically late project, a huge performance regression, or a scary incident—and you could save the day. Once again, it feels nice to be needed! And it can be oddly relaxing to join in on a crisis: the goals are usually very clear and there's a bias toward action rather than consensus and planning. But it's an abrupt transition.

If there's a sudden crisis that calls for all hands on deck (or *you* on deck), you might be abruptly doing something else for a while, then returning to your regular project schedule (as in Figure 4-9). It's a major context switch. It can be a bit jarring, and afterward it might take you some time to get back on track with whatever you were doing before. But helping out is often the right thing to do.

Remember, though, that if you do *too* much crisis response, it can be hard to find opportunities for growth, or to have much of a narrative for your work other than "I jumped on whatever the current fire was."

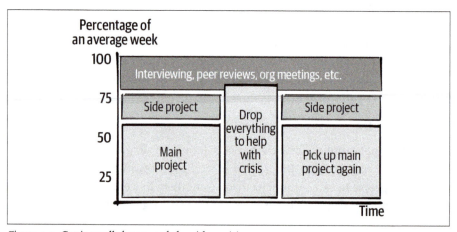

Figure 4-9. Getting pulled away to help with a crisis.

You're claiming a problem

After the fifth time you're slowed down by a clunky API or some wodge of technical debt, you might find it hard to not just go take care of it. If you ranked all available work, you might be hard-pressed to claim that this fix is the highest on the list, but it's *bugging* you and you don't want to ignore it any longer. Time for a side quest (see Figure 4-10) to fix it!

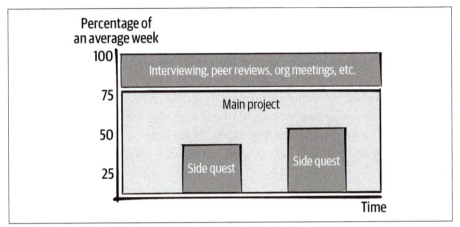

Figure 4-10. Side quests take time away from your intended work.

When you're blocked on bigger work, solving small, easy problems can be a great way to build up momentum again. You get to fix the problems in front of you, and probably get a lot of kudos (and credibility and social capital!) from your peers. It feels amazing. But too much of this kind of work can keep you away from longer-term projects.

You're invited to join a grassroots effort

A more nebulous invitation is to get involved with some grassroots initiative, say, a working group to improve testing across the company. This sort of initiative is not on anyone's OKRs and probably not something the company is already invested in. Joining a group like this can be a way to work with interesting people and have a huge impact, or it can be a *tremendous* waste of time. It can be hard to tell which you're heading into, so be wary. Working groups can be effective if there's organizational buy-in, a clear time commitment, exit criteria, and a process for making decisions. But if it's a big group of overallocated people who like to talk but have no power to change the things that are bothering them, it's a social group and no work will occur.

Someone needs to...

It could be action items at the end of a meeting, an unexpected request from another team, or a problem that nobody anticipated: some work doesn't belong to anyone, and you should do your fair share. If it will involve horrible levels of politicking, cruft, or tedium, consider taking on *more than* your fair share: sometimes being the most senior person around means that you should shield your less experienced colleagues and take on the grungiest work. Volunteering can build credibility and goodwill with your team, and while it's often frustrating, there's a weird satisfaction in getting to the end of the work and knowing that you're the reason it's no longer a mess. But just like with jumping on problems, watch out for doing too much of it at the expense of other results people are expecting from you.

You're just meddling

As a staff engineer, you'll likely have opinions on a bunch of things (as in Figure 4-11) you haven't been invited to help with. If you see a project that's going in a dangerous direction, or you have ideas about how to make it better, you might nudge your way in and have some conversations. Meddling can be very welcome, or very much the opposite. Make sure you're not "lobbing a water balloon": getting involved long enough to cause chaos, then disengaging without sticking around to experience the consequences of your changes.

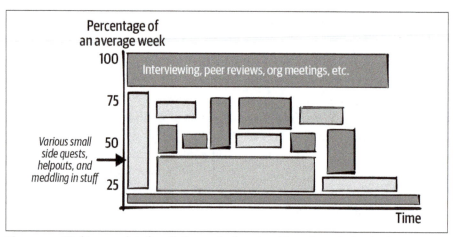

Figure 4-11. Working on a lot of small, frequently changing things. It can be hard to have a narrative for this kind of work.

Other People Aren't Necessarily Right

Some of us allow low-priority other-people work to preempt even the most high-priority things we were planning to do. You might be one of those people if you always drop your own work to help with even a minor outage, or respond on Slack within seconds, no matter what else you were doing.

Sometimes interruptions really are more important than what you were intending to do. But be discerning. Remember that other people have limited access to your resource dashboard. Unless you're very lucky, only you will be taking care of your own resources.

WHAT ARE YOU SIGNING ON FOR?

As you consider taking on a new project, task, mentoring arrangement, meeting, etc., understand its shape and ask yourself what impact it will have on your time, both now and later. Be clear with yourself about what you're adding to your schedule and sending to Future You. Some projects start small but hit an inflection point (such as project approval), then start needing much more time. Others are time-intensive at the start and then trail off. If you know it's going to get busy later on (as in Figure 4-12), will you still be able to handle it?

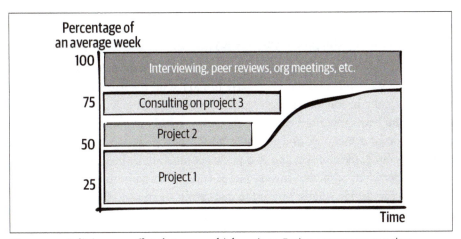

Figure 4-12. Splitting your efforts between multiple projects. Project 1 ramps up over time.

Be cautious about how you estimate time. If you're agreeing to a one-hour meeting every week, for example, is that really an hour, or will you need to do extra prep and come away with action items?[14] If the meeting is in the middle of an empty four-hour block, will you still be able to use the other three hours productively?

A project that seems small can actually be a huge time commitment if you're going to stick with it for a long time. Consider the boundaries and scope of your involvement. If you're taking on ownership of a process, will you be able to hand it off to a team later on? If you're temporarily joining a project or helping out with a crisis, but don't intend to be on it long term, be clear that you're joining just for the beginning to disambiguate it or to help the lead get started. Exit criteria are especially important for work that is risky or that you're not sure you should be doing at all. You can reduce that risk by including formal go/no-go or retrospective points, or framing things explicitly as a timeboxed experiment.

Even small projects or tasks can affect your resources, and their effect can be disproportionate. A single meeting can build your social capital and credibility substantially *or* drain your energy and morale so that you can't focus for the rest of the day. Pairing with someone very skilled for an afternoon can sometimes teach you more than you'd pick up in a week of learning on your own.

QUESTIONS TO ASK YOURSELF ABOUT PROJECTS

Here are a few more resource questions to think about.

Energy: How many things are you already doing?

I joke that I have enough energy to care about five things at once.[15] If I choose to care about a sixth, one of the previous five has to fall off the list. Otherwise, as the number of balls in the air increases, I'm likely to accidentally drop one—and it won't be the least important one. I can usually care about some extra things for the duration of a meeting, and maybe a brief action item or two afterward. But after that's done, the slot is going to quickly get filled up by the topics in the next meeting. So when someone invites me to care about some new problem they have, I have to decide whether I want to *stop* caring about something else.

If the new project is asking you to care about something new, do you have a free slot for that? Remember that your capacity will be affected by what's

---

14 Budget another few hours if you're a chronic volunteer.

15 OK, only mostly a joke.

happening in "real life" too, not just work. If you're expecting some huge life event, this is probably not the time to start a project that will need a ton of your energy and attention. Every choice means there's something else you can't do, and if you choose too many things, you end up diluting your impact on any one of them. So you really do need to choose your battles.[16]

Warning

If you tend to let distractions pull your attention in lots of directions at once, try to build the habit of pausing for a few seconds before reflexively volunteering or agreeing to do something.

Energy: Does this kind of work give or take energy?

What kind of work will this project involve? If it's going to take a gazillion whiteboard conversations to build up context and you find those draining, this project is going to be more expensive for you. If you find it hard to focus for a long time and you're going to need to read hundreds of pages of backstory or a stack of industry white papers, that kind of project will be harder for you.

Do any of the people you'll work with leave you exhausted every time you talk to them? If you're considering a project and get tired just thinking about the kind of work that's involved, weigh that up when you're deciding whether to start it.

Energy: Are you procrastinating?

If you're low on energy, it can be almost impossible to push through and take the next step on some huge, ambiguous, complex project. You might find yourself picking up some low-priority work to fill the gap. That can be OK: the last hour of the day after some draining meetings can be a great time to tidy your desk or archive old mail. But some low-priority tasks can start out easy, then grow into another complicated project that you'll need energy to tackle.

In his blog post "The first rule of prioritization: No snacking" (*https://oreil.ly/x5803*), Intercom cofounder Des Traynor discusses the magnetic pull that low-effort, low-impact work can have for engineering teams. He describes Hunter Walk, now a partner at Homebrew, drawing a 2 x 2 graph to map impact against effort (see Figure 4-13), and warns against the quadrant of low effort and low impact, describing such work as "snacking." Since this work tends to be quick

---

16 As the philosopher Ron Swanson (*https://oreil.ly/VEFBv*) tells us, it's better to whole-ass one thing than half-ass two things.

and useful (and feels good), it can be easy to justify doing a lot of it. But, as Traynor notes, "It feels rewarding and can solve a short-term problem, but if you never eat anything of substance you'll suffer."

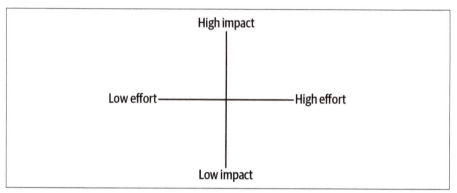

Figure 4-13. Projects can be high or low impact, high or low effort. Be wary of spending too much time in the low-impact, low-effort quadrant.

When you're tired, snacking can feel easier than resting. Notice when you're doing busywork because you're tired, and find a way to rest instead.

Energy: Is this fight worth it?

In her article "OPP (Other People's Problems)" (*https://oreil.ly/YoE84*), Camille Fournier cautions about the grinding frustration of trying to solve all of the unowned problems you see. Even if it ends successfully, an organizational project can take longer and drain more energy than you'd intended. As Fournier writes: "Take a moment to reflect on whether it was worth the effort to you, and think about how many more things like this you see at the company you're in that you really want to change just as much as that one. Think about what else you could be doing with that extra energy."

Quality of life: Do you enjoy this work?

Think about the kind of work that your project will need. In Chapter 1, I quoted Yonatan Zunger's observation that every project needs core technical skills, project management, product management, and people management. Depending on who else is on your project, some of those categories may not be assigned to anyone and, if you're leading the project, you'll be filling in the gaps. Will you enjoy that kind of work, or will you resent it for not being the thing you actually *want* to do?

Here are some more points to consider:

- If you like your work to be intense and have high stakes, will the project have enough excitement to keep you happy? Conversely, if you prefer things to be a little more predictable, will it be too stressful?
- If you hate working alone, will you have people to collaborate with? If you find people exhausting, will you need to be in a big group all day?
- If you like pair programming, will this project let you pair? If you hate pair programming, will this force you to do it?
- Will you have to be on call? Do you like being on call?
- Will the project require travel? Do you want that?
- Will it provide opportunities for conference talks, writing articles, or otherwise publicly sharing what you learn? Do you want those?
- Will you get to work with people who make you feel comfortable and safe? Will you be able to relax and be yourself around your coworkers?

Quality of life: How do you feel about the project's goals?

Will this work align with your values and make you feel fulfilled, or do you feel a bit "off" about it? For some people, value-aligned work means making sure you're not doing something that hurts others. For others, your work needs to actively help create a better world. What is your project aiming to do? In most cases, changing projects inside a company will not change how much your work aligns with your values, but sometimes you'll be doing something you feel better or worse about. Think about the positive or negative effects of your work on the world, and weigh up how it affects your own personal satisfaction with your life.

Credibility: Does this project use your technical skills?

If you've been operating at a very high altitude for a while, you might want to get into the trenches occasionally to show that you still know what you're talking about. Different projects will use and showcase different skills, and doing difficult things gives you more credibility. If you can implement something that three other people have already failed at, or make it tractable for other people, that's a solid boost to your reputation—and if you later want to do something more abstract, you'll have less risk of being seen as living in an ivory tower.

Credibility: Does this project show your leadership skills?

A new project can be an opportunity to show that you're a leader through taking responsibility for outcomes, communicating frequently and well, and giving the right level of detail around what's going well and what's not and why. And of course, you'll build trust with your organization by actually succeeding at the project. Evaluate new projects for what kinds of skills they're going to let you demonstrate and how succeeding will reflect on you. Also be aware of the risk of failure: if you're taking on an unwinnable battle, will you be respected for trying, or will you be blamed?

Social capital: Is this the kind of work that your company and your manager expects at your level?

In Chapter 1, we explored what you and your organization believe your job is. You may want to optimize for the kind of work your manager considers appropriate for your level (or for a level you'd like to get promoted to). If you're somewhere a staff engineer isn't considered a success unless they're writing code (or designing systems or leading big projects), make sure you're doing those things.

In general, work that matters to the people in your reporting chain is work that builds social capital. Lest this start to feel *really* Machiavellian, I want to reiterate that this is just one aspect of the project! I suspect we all know the kinds of people who *only* optimize for looking good to leadership, and those aren't people we tend to respect. But do keep an eye on your current standing with the people who influence your calibration, compensation, access to good projects, and future promotions. Managing up (*https://oreil.ly/AILvi*) includes understanding your boss's priorities, giving them the information they need, and solving the problems that are in their way—in other words, helping them be successful.[17] Their success gives *them* social capital that they can spend to help you.

Social capital: Will this work be respected?

Your project can build—or lose—social capital with your peers depending on how it aligns with their values. If you're working on something that other people consider to be an important fight, or just a very cool project, that will build goodwill and they'll be more inclined to help you. If they think you're doing something pointless, misguided, or even evil, they're going to think less of you, and

---

17 If you're more senior than your manager and "reporting low" (see Chapter 1), managing up might also mean coaching them and helping them grow into their role.

you'll struggle to get their assistance or trust. I've seen many disapproving back-channel conversations along the lines of "I'm really surprised to see [person] work for [distasteful project or company]." If that person is well regarded, they might be trusting their endorsement to add a shine to a shady-looking project. Sometimes they're right, but attaching their reputation to the project is a risky move.

Social capital: Are you squandering the capital you've built?

Here's a cautionary tale. I once worked on a team that hired a new senior-level engineer, someone who came with big credentials and a lot of respect. It was widely acknowledged that he'd been the only reason his previous project had succeeded, and we had high hopes for what he would do. And he started great! As the most senior engineer on the team, he was immediately productive, solving big problems and raising the standards for everyone else. We were so glad he'd joined—until we weren't.

After just a few weeks on the team, he noticed a system that wasn't following a best practice. He wasn't wrong—it was a mess—but it wasn't a system we touched often, and it didn't seem worth fixing. Our new colleague advocated for the change anyway. His reputation and his enthusiasm swayed everyone else, and nobody objected when he took two new grad engineers away from their projects and dove headfirst into this one. Bad call. After several weeks, it became clear that this project would take a few quarters—and cost 10 times what he'd anticipated. But the senior engineer kept pushing ahead, insisting that it would be worth it. It was a couple of months before he accepted reality and the project was abandoned. The new grads returned to their previous priorities, and everything went back to normal—except that our new hire was no longer quite so esteemed. He'd squandered his social capital on a fight he didn't particularly care about. What a waste.

Notice when you're in danger of wasting your social capital. And be deliberate when you spend it to help other people, such as asking your company to interview a friend who has a résumé they would normally pass up. This form of support, sometimes called *sponsorship*, costs you something: if the person ends up failing, being a jerk, or otherwise being a regrettable choice, it will reflect on your judgment. Make sure the person you're sponsoring is worth it.[18] Be aware of

---

18 As we'll discuss in Chapter 8, it's really easy to accidentally find yourself only sponsoring people like yourself. Watch out for implicit bias here.

this dynamic too when you're looking to *borrow* capital from someone else—for example, when you want executive sponsorship for a technical vision or strategy. If you're borrowing someone else's authority and reputation to move a project along, don't squander it. They might not sponsor your ideas a second time.

Skills: Will this project teach you something you want to learn?

Tech changes fast, and your skills will become out of date over time unless you're deliberate about keeping up. A new project can be a great opportunity to practice a skill you want to get better at. This could be for a role you're aspiring to, aspects of your current role, or just topics you'd enjoy knowing more about.

One way to think of this: what stories do you want to be able to tell on your future résumé? Do you want to show that you can take on a big, ambiguous, messy project and make it happen? Do you want an example of debugging something difficult, driving a major culture change, or turning junior engineers into senior engineers? If so, look for projects that give you opportunities to practice those skills.

Skills: Will the people around you raise your game?

Some people make you better at your job without setting out to teach you anything: they're so competent that you build skills just by bathing in their aura. OK, the reality is that you learn by watching the skill being executed well (*https:// en.wikipedia.org/wiki/Observational_learning*), but I've definitely had colleagues who seemed to have a magical effect on their team. A whole lot of Part III of this book is going to be about *being* one of those people. But I recommend you try to find people like that for yourself too.

Even if you're the most senior person in your group, you can still learn from the people around you. People who do great work tend to elevate the skills even of those more senior than them. And if you have a team where the new grads are in awe of the skills of the seniors, but the seniors are just as much in awe of the skills of the new grads, that's a team where everyone makes everyone else better.

Working with someone who's great at a skill you want can take you up a level in a way that's hard to find otherwise. This is why internships can be so valuable. But the same phenomenon holds throughout our careers. As a friend said when she was offered an opportunity to work with her company's CTO: "I'll get to learn how CTOs talk to people." So, when you're choosing a project, look at whether you'll work with people you'll learn something from, and people who will inspire you to do your best work.

WHAT IF IT'S THE WRONG PROJECT?

After viewing a potential project through the lens of each of those questions, you've probably got a good feeling for whether you should take it on. Maybe you shouldn't! It's possible for a project to be important but not actually right for you.

If you've decided that a project isn't a good fit for your current schedule and needs, you have a few options. You can try to do the thing anyway and accept the consequences: a popular choice, but not a long-term sustainable one. You can try to compensate for the negatives by canceling other work to make space on your time graph, or getting your needs met elsewhere. You can make the project into an opportunity for someone else. You can reshape it so that it does fit. Or you can just say no. Let's look at each of those.

Do it anyway?

A popular approach for projects that don't fit is to nonetheless try to make them happen. It feels like the path of least resistance: you don't have to say no, and surely (you tell yourself) Future You will figure something out.

This decision can even be the right one in the short term: you might take one for the team and get a vital project done, even if it's terrible for your time, energy, skills growth, etc. If you find yourself taking on work that undermines your needs, though, make sure you understand why. Is this a temporary situation? What are the exit criteria and when do you expect to get there? If the plan relies on self-sacrifice, that's not something you can or should enable long term.

If you've been doing projects that aren't good for you and there's no end in sight, this is a good conversation to have with your manager. If you tell them "This work is sapping my energy and spilling over into my family time" or "I want to do something higher-profile because I'd like to get promoted" or even "This project is just making me really unhappy," they should listen. Hiring staff engineers is difficult and expensive, and most good managers will have a little alarm signal going off in their brains if a senior engineer is emoting unhappiness with their work. That doesn't mean the situation will change immediately, but your manager should help you find a path over time to something more compatible. If you have been very clear about your needs and are still unable to make a change, that might be a signal that you're in the wrong team or the wrong company.

This realization does not necessarily reflect badly on anyone: you might have become a big fish that needs to move to a bigger pond. I've seen this happen a lot for teams working on systems or products with relatively simple needs or small scope: what the individual needs and what the team needs stop aligning. It feels harsh to say "If you want to grow, it's time to do something else," but that's often the reality.

Compensate for the project

If a project is a good fit in some aspects but not in others, is there a way you can compensate for it? For example, can you add a side project that gives you the enjoyment, skills growth, or credibility that you're not getting from your main project? If the gap is that you just don't have time and energy for a particular project, can you make space? If you're using the trick of putting your work into your calendar, you get a nice visual representation of this. If you know something will take ten hours and you're having trouble scheduling time for it, you'll probably need to move something else out of the way.

And if you've got projects taking up space in your brain that need more time and attention than you can give, it's OK to decide that you're going to stop caring about one of them. Maybe you're mentally keeping several projects warm on the back burner, hoping that someday you'll have time to get to them. My colleague Grace Vigeant gave me great advice once: sometimes you have to torch the back burner.[19] Accept that a task is not important enough to get back to—or hand it over to another chef!

Let others lead

A project that isn't a good fit for you right now might be an excellent opportunity for someone else. Think about your coworkers' energy, skills, quality of life, credibility, and social capital, and see if there's someone else who will benefit from the project. Author and engineering leader Michael Lopp says that (https://oreil.ly/sVogl) a leader's job is to "aggressively delegate." He says that there's guaranteed to be work that shows up on your plate on which you can "get an A" every single time, and if you give it to someone else, they're probably going to get a B. But, he argues, a B is a pretty great outcome for their first time doing this kind of work: "You're demonstrating trust by giving them work that's scary to them and that you know—and they know—is beyond their means. 'I know you can do this. I'm

---

19 If you're using the trick of putting work in your calendar, it can feel very freeing to get to that "meeting" and decide not to do it or even reschedule it. You're not going to do it. It's *gone*. I'm told that bullet journaling is helpful for letting go of things, because every day you copy the things you still care about to the next page. You can decide not to move something, and just…let it go.

going to help you with this.' That's amazing." The other person gets to learn. And maybe you can coach them from a B to an A.

If you're the most senior engineer in the group, make sure you're not taking all of the opportunities to be publicly competent. Do a gut check for whether someone else needs the project more.

Resize the project

If the project doesn't work for you in its current form, sometimes you can reshape it into something that does. Maybe you can't join the project full time, but you can join for the first month to evaluate the direction for feasibility, or act as a consultant to a different lead. If you don't have free cycles to take on an ongoing mentoring relationship, maybe you can meet once. If you aren't available to review a proposed design, maybe you can recommend someone else. And if you aren't willing to let a certain problem become one of the things you're going to care about on an ongoing basis, can you care about it for the duration of a meeting and offer advice on how you'd proceed?

If a project would be interesting to you with some modifications, it's usually worth talking about that. The worst they can say is no.

Just don't do it

Of course, the final option is to *just not do the thing*. Oof, it's easier said than done! It can be hard to leave something broken, to notice a problem that you know you could solve, and ignore it every day. Sometimes you have to, though. Not all problems are your problem. Either you'll get to it later, or someone else will, or it will stay broken and that just gets to be OK.

It can be even harder to say no if someone has asked you for help, especially if you really could have stretched and done the thing. But saying no is the price of high-quality work: if you do too many things, you won't be able to do them well.

It's common to feel uncomfortable saying no, enough that there are a ton of articles out there with scripts for how to do it. I like this one from indiatoday.in (*https://oreil.ly/RuzbG*), for example, which recommends "I wish there were two of me," "Unfortunately, now is not a good time," and "Thank you so much for thinking of me, but I can't!" Or this excellent advice (*https://oreil.ly/KbzSV*) from the *Ask a Manager* blog:

Pay attention to how people you admire say no. You might be wary of pushing back on a request because you can't imagine how to do it in a way that doesn't alienate people. Look at colleagues who seem to do it successfully, and see if you can find language, tone, and other cues that you can adopt for yourself.

Or just think back on the times you said yes and wished you hadn't! Your "no" is a gift to your future self. I've started tagging emails with an #isaidno label in Gmail, an idea I got when Amy Nguyen tweeted about doing something similar (*https://oreil.ly/6SR5Q*). It feels *so good* when the deadline rolls around and I'm not on the hook.

EXAMPLES

Let's end this chapter by looking at some examples, weighing up the costs and benefits of each, and working through ways to reduce the cost and increase the benefits. I'm going to suggest some outcomes, but of course you might make different trade-offs.

Example: Speaking at the all-hands meeting

You've just shipped a project that's been running for 18 months. It was a difficult project, and you've been working flat-out for the last couple of months. It shipped, it landed well with customers, and you're feeling triumphant. Also *tired*. You're about to click Submit on the PTO form when you notice an email from your VP. They'd like you to present about the project at the engineering all-hands in two weeks. You'll be back from vacation then, but it'll mean creating a presentation deck when you'd planned to be on a beach. What do you do?

Let's weigh up this opportunity, illustrated in Figure 4-14. It's great visibility, and your VP is inviting you to do it. That's a boost to credibility and social capital. It'll take time, though: you won't want to turn up with a half-assed presentation, so it's going to eat into your vacation time, denting your quality of life. It'll take a lot of energy, and you're very tired after the big project. What should you do?

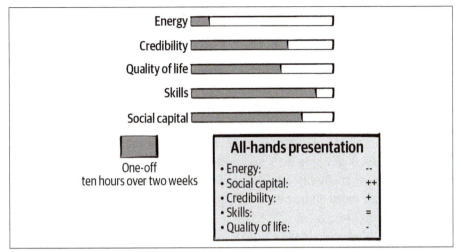

Figure 4-14. Extra social capital and credibility, at the cost of some quality of life and a lot of energy you don't have to spare.

The biggest question I'd ask you is whether this is a *new* opportunity for you. If you're used to presenting to groups this big, you're not going to learn as much or get as much benefit from this opportunity as if it's your first time ever. Are you coming from a deficit of social capital or credibility? Do you need this boost, or does launching the project mean that you're already well regarded?

Unless this is an amazing opportunity for you or you really need this particular win, I'd look at who else this could be an opportunity for. Was there someone else on the project who did good work but hasn't had as much glory from it, or who will learn a lot by presenting? Make sure you pick someone who will put in the work and do a good job. It doesn't have to be as good a job as *you* would have done, but if you spend social capital to recommend someone who shows up with shoddy slides they created in 10 minutes, it will reflect badly on you.

If you decide that this is an opportunity you should take, it might be worth trying to reshape this project. Can you make it a shorter presentation and lower the effort? Can you copresent with someone and have them take some of the load of the slides? Best of all, can you postpone to the following all-hands meeting? By then, you'll have more information about usage numbers for your feature.

Example: Joining an on-call rotation

Your company has an on-call rotation for incident commanders: the people who step up to coordinate major incidents that are visible to users or cross multiple teams. You haven't been an incident commander before, and you like the idea:

it's an opportunity to work across teams in a very visible leadership role, as well as get into some interesting technical problems in real time. Plus, you never learn as much about your systems as you do when something is broken. It's a little scary, though, because if there's a big outage, all eyes will be on you. There'll be a learning curve that will take time. And, of course, you'll be on call and can be paged out of hours. What do you do?

Figure 4-15 shows how your resources will change if you take on this role. Since you haven't been an incident commander before, this is a skill boost for a very transferable skill. Being the responsible person during a major incident also tends to increase credibility and social capital—assuming you do it well. But doing it well will mean an investment of time. And being paged out of hours can be rough on your energy.

I'd weigh this one up based on quality of life. How resilient you are feeling at this stage of your life? If you're feeling a bit fried, adding an occasional extra wake-up could cost more than you should spend right now. If you've got a young child or your mental health needs managing, then it might be the wrong time in your life to build a skill that comes with interrupted sleep.

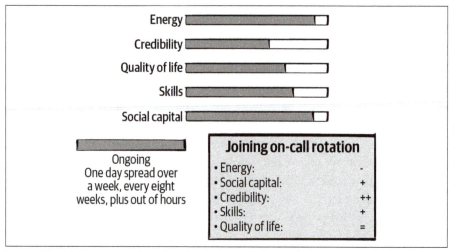

Figure 4-15. Joining an on-call rotation is a fairly small-time commitment most of the time, but can be a huge energy drain.

On the other hand, if you've recently been working on high-level projects with long feedback loops, it can be *fun* to do something where you can see your impact immediately. As one manager friend says, "At the end of an incident, coming down from the adrenaline, it's really clear what I did today." If the on-call rotation comes with extra compensation, is it enough to compensate for the negatives? Do it if the idea of doing it makes you happy. Maybe sign up for six months and then reconsider.

Example: The exciting project you wish you could do

You've been at a company for a few years and you've done a lot of work to modernize architecture and processes. As a result, you're now the point of contact for a lot of things. If there's a question about how the company does testing, onboarding, incident response, or production readiness, you'll end up in a meeting about it. You've got a long backlog of improvements you'd like to make to these processes, including some that are underway.

There's a new project coming along in an area you don't know a lot about but find really interesting. Given an empty calendar, you could learn the base technology, and you know you could lead the project and make it successful. But you don't have an empty calendar! You're worried that your lack of time would put the project at risk. But you'd love to do this work and you'd love to *have done* it too. It'd make a great résumé line. What do you do?

Figure 4-16 describes the effect on your resources. It sounds like this is a project that would make you happy and build skills that you want to have. It might be a boost to credibility, too. But if you take it on and fail, that'll look pretty bad for you. And it is likely to take more time than you have available, so failure is a real risk.

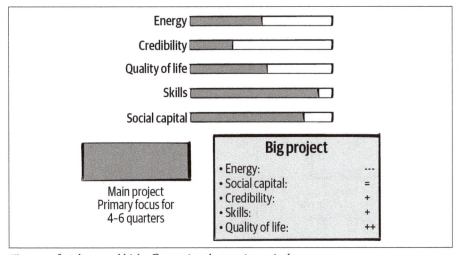

Figure 4-16. A huge and high-effort project that you're excited to start.

What should you do? It depends on whether you're ready to make space and torch the back burner. This is a good conversation to have with your manager, or their manager: how important is it that someone keeps iterating on the existing processes? Are you polishing something that's pretty good already? If the work is needed, is there someone you can hand it off to?

I think you should take the new project. Be careful about drifting back into the process work, though. This is a time when putting nonmeetings in your calendar will be crucial: if your calendar is defined by what other people put into it, you're not going to have time left over for your project. Block out the time you need for the project first, and then let other people claim time from what's left.

Example: I want to want to

Your manager has asked you to lead a project. It's critical to the business, highly visible, and the sort of thing you could do really well. It's bigger than anything you've done before, and your coworkers would be a dream team of people you'd enjoy working with and learning from. There's only one problem: you *just don't want to*. It's frustrating, because on paper the project looks amazing for your career, a good opportunity that you may regret passing up. You wish you wanted to do it, but you just don't. The project needs the kind of work that you've done a lot but don't enjoy, and you're longing to do something else—something you'll do less well but that will make you want to come to work in the morning. Figure 4-17 shows the situation. What do you do?

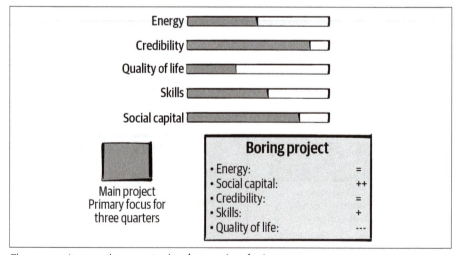

Figure 4-17. An amazing opportunity...but you just don't want to.

If you feel that strongly negative about anything, pay attention to the feeling; don't rationalize it away. No matter how good the opportunity, taking it isn't mandatory.

Before you say no, consider whether there are ways you could reshape this project to fit your needs better. Would it be more interesting to you to coach someone else as lead, or join in a different role that's still helpful for your manager? Would you be interested in joining for a month to get the project moving? It's OK if you still don't want to, but think through whether there's a variant that would work better. Finally, if you're looking for a different kind of work, tell your manager and other people in your network, so they'll think of you when something new comes up.

DEFEND YOUR TIME

A few years ago, I was considering emailing a VP of engineering to ask for advice. I'd briefly worked with her years earlier and I was sure she'd remember me, but the problem I was trying to solve didn't have anything to do with her or her organization. She was just someone I knew who would have a helpful perspective. I cast around for less busy people I could ask, but nobody came to mind. So I worried about it out loud with friends on a group chat: "I want to ask her, but I know she's busy. But it would really help. But I don't want her to feel under *obligation* to help me." One friend gave me advice that I'm still using years later: "Ask her. You don't get to that level without knowing how to defend your time."

You won't succeed unless you can defend your time. The number of demands on it will increase and the number of available hours will stay the same, so be deliberate about what you prioritize.[20] And, when you choose a project, make sure you have enough of the resources you'll need to do a good job.

In the next chapter, we're going to look at leading big projects.

---

20 The available hours may even decrease. As Will Larson says in one of my favorite of his articles, "Work on What Matters" (*https://oreil.ly/OQW3K*), "Even for the most career-focused, your life will be filled by many things beyond work: supporting your family, children, exercise, being a mentor and a mentee, hobbies, and so the list goes on. This is the sign of a rich life, but one side-effect is that time to do your work will become increasingly scarce as you get deeper into your career."

To Recap

- By the time you reach the staff+ level, you will be largely (but probably not entirely) responsible for choosing your own work. This includes deciding on the extent and duration of your involvement with any given project. You can't do everything, so you'll need to choose your battles.

- You are responsible for choosing work that aligns with your life and career needs as well as advancing your company's goals.

- You are responsible for managing your energy.

- Some projects will make you happier than others, or improve your quality of life more.

- Your social capital and peer credibility are "bank accounts" that you can pay into and spend from. You can lend credibility and social capital to other people, but be conscious of whom you're lending to.

- Skills come from taking on projects and from deliberately learning. Make sure you're building the skills you want to have.

- Free up your resources by giving other people opportunities to grow, including starting projects and handing them off.

- Focus means sometimes saying no. Learn how.

5

Leading Big Projects

What makes a great project lead? It's rarely genius: it's perseverance, courage, and a willingness to talk to other people. Sure, there might be times when you need to come up with a brilliant and inspired solution. But, usually, the reason a project is difficult isn't that you're pushing the boundaries of technology, it's that you're dealing with ambiguity: unclear direction; messy, complicated humans; or legacy systems whose behavior you can't predict. When the project involves a lot of teams; has big, risky decisions along the way; or is just messy and confusing, it needs a technical lead who will stick with it and trust that the problems can be solved, and who can handle the complexity. That's often a staff engineer.

The Life of a Project

In this chapter we're going to look at the life of a big, difficult project. While projects come in many shapes, as we saw in Chapter 4, I'm going to focus on the kind that lasts for at least several months and needs work from multiple teams. For the purposes of this chapter, I'll assume that you're the named technical lead of the project, perhaps delegating to some subleads of smaller parts. I'll assume that nobody who is working on the project is reporting to you, but that you're nonetheless expected to get results. There are probably other leaders involved: you might have project manager or product manager counterparts, and there could be some engineering managers, each of whom has a team working on their own areas.[1] But, as the lead, *you're* responsible for the result. That means

---

1 I'm going to say "project manager" throughout this chapter, but you might work with a program manager (usually someone who's responsible for multiple projects that have a shared goal). Sometimes you'll see the title technical program manager (TPM). For the purposes of this chapter, just assume I'm talking about someone in any of these roles, and that they're preternaturally organized and competent at delivering products.

you're thinking about the *whole* problem, including the parts of it that lie in the fissures between the teams and the parts that aren't really anyone's job.[2]

We'll begin before the project even officially starts, when you're looking at a vast, unmapped, and quite possibly overwhelming set of things to do. We'll work through some techniques for getting the lay of the land and making sense of it all. And we'll talk about creating the kinds of relationships where you're sharing information and helping each other rather than competing.

Then we'll set this thing up for success the way a project manager would: thinking through deliverables and milestones, setting expectations (including your own), defining your goals, adding accountability and structure, defining roles, and—the number one tool for success—*writing things down.*

After the project is set up and humming along, we'll look at *driving* it. You've got a destination, and you'll need to make some turns and course corrections to get there. I'll talk about exploring the solution space, including framing the work, breaking the problem down, and building mental models around it. When a project is too big for any one person to track all of the details, narrative is vital. We'll look at some common pitfalls you might meet during design, coding, and making big decisions. The chapter ends on spotting the obstacles in your path— the conflict, misalignment, or changes in destination, and how to communicate clearly while you navigate around them.

We won't look at the end of the project until Chapter 6, but for now, let's start at the very beginning.

The Start of a Project

The beginning of a project can be chaotic, as people mill around trying to figure out what they're all doing and who's in charge. Congratulations: as the technical lead, you're in charge. Sort of.

The other people on the project aren't your direct reports, and they're still getting instructions from their managers. You have...maybe?...a mandate to get something done. But it's possible that not everyone agrees yet on what that mandate *is* or whether they're supposed to be helping you with it. If several different managers or directors are involved, it might not be clear what you're responsible for and what they're expecting to own. There might be other senior engineers on

---

2 If you skipped Chapter 2, just imagine teams and organizations as tectonic plates moving against each other, with friction and instability where the plates meet.

the project, maybe some more senior than you. Are they supposed to follow your lead? Do you have to take their advice?

IF YOU'RE FEELING OVERWHELMED...

Maybe you're joining an existing project, with all of its history and decisions and personality dynamics and documentation. Maybe the project is new, but there are already detailed requirements, a project spec, milestones, and a documented list of eager stakeholders. Or maybe there's just a whiteboard scrawl, or—frustratingly often—a bunch of long email threads (some of which you weren't cc'd on) that culminated in a director deciding to fund a project to solve a poorly articulated, unscoped problem. Almost certainly there are other people who want to give you their opinions on that problem, and there might be immediate deadlines you want to get ahead of. All of this comes up before you really have a handle on what the project is for, what your role is, and whether you all agree on what you're trying to achieve. It's a lot to think about.

What do you do? Where do you even *start?* Start with the overwhelm.

It's normal to feel overwhelmed when you're beginning a project. It takes time and energy to build the mental maps that let you navigate it all, and at the start of the project it might feel like more than you can handle. But, in the words of my friend Polina Giralt, "that feeling of discomfort is called learning" (*https://oreil.ly/2qXQH*). Managing the discomfort is a skill you can learn.

You might even find yourself feeling that you've been put in this position by mistake or that the project is too hard for you, struggling with fear that you'll let others down or fail publicly: a common phenomenon known as *imposter syndrome.* Emotional overwhelm can get in the way of absorbing knowledge and even affect your performance, making impostor syndrome almost self-fulfilling.

These feelings might be a signal that you're low on one of the resources from Chapter 4. If you've exhausted all of your energy, you're low on time, or you don't feel like you have the skills to do what you need to do, that may manifest as stress and anxiety. Check in with yourself and ask whether any of your resources are at worrying levels.[3] Is there anything you can do to get more time or energy or build more skills? Are there people who could help?

---

3 Check in on your biological needs too. Not trying to get in your business, but lots of people in our industry are sleep-deprived. Are you? Sleep builds resilience and willpower and energy! It's amazing. And when did you last drink a glass of water?

You might also think about how this work would feel if someone else was doing it. George Mauer, director of engineering at findhelp.org, told me that he used to feel imposter syndrome, until he realized "99% of people don't know better than I what to do." Maybe you're figuring out what you're doing as you go along, but hey, everyone else is too! Is it just me, or is that *really* reassuring? No matter who was doing this project, they'd find it difficult too.

The difficulty is the point. I find that I can handle ambiguity when I internalize that this is the *nature of the work*. If it wasn't messy and difficult, they wouldn't need you. So, yes, you're doing something hard here and you might make mistakes, but someone has to. The job here is to be the person brave enough to make—and own—the mistakes. You wouldn't have gotten to this point in your career without credibility and social capital. A mistake will not destroy you. Ten mistakes will not destroy you. In fact, mistakes are how we learn. This is going to be OK.

Here are five things you can do to make a new project a little less overwhelming:

Create an anchor for yourself

Here's how I start, no matter the size of the project: I create a document, just for me, that's going to act as an external part of my brain for the duration of the project. It's going to be full of uncertainty and rumors, leads to follow, reminders, bullet points, to-dos, and lists. When I'm not sure what to do next, I'll return to that document and look at what Past Me thought was important. Putting absolutely everything in one place at least removes the "Where did I write that down?" problem.

Talk to your project sponsor

Understand who's sponsoring this project and what they'll want you to do for them. Then get some time with them. Go in prepared with a clear (ideally, written) description of what *you* think they're hoping to achieve from the project and what success looks like. Ask them if they agree. If they don't, or if there's any ambiguity at all, write down what they're telling you and double-check that you got it right. It's surprisingly easy to misunderstand the mission, especially at the start of a project, and a conversation with your project sponsor can confirm that you're on the right path (which is always reassuring). This is also a good time to clear up any confusion about what your role will be and who you should bring project updates to.

Depending on the project sponsor, you might have regular access to them, or you might get a single conversation and then nothing more for months (a horrible way to work, but it does happen). The less often you're going to talk with them, the more vital it is that you get all of the information up front.

Decide who gets your uncertainty

Think about who you're going to talk with when the project is difficult and you're feeling out of your depth. Your junior engineers are not the right people! While you can and should be open with them about some of the difficulties ahead, they're looking to you for safety and stability. Yes, you should show your less seasoned colleagues that senior people are learning too, but don't let your fears spill onto them. Part of your job will be to remove stress for them, making this a project that will give them quality of life, skills, energy, credibility, and social capital.

That doesn't mean you should carry your worries alone. Try to find at least one person who you can be open and unsure with. This might be your manager, a mentor, or a peer: the staff engineer peers I discussed in Chapter 2 can be perfect here. Choose a sounding board who will listen, validate, and say "Yes, this stuff is hard for me too" rather than refusing to ever admit weakness or just trying to solve your problems for you. And, of course, be that person for them or others too.

Give yourself a win

If the problem is still too big, aim to take a step, any step, that helps you exert some control over it. Talk to someone. Draw a picture. Create a document. Describe the problem to someone else. In some ways, the start of the project is when it's easiest to not know things. You can preface any statement with "I'm new to this, so tell me if I have this wrong, but here's what I think we're doing" and learn a lot. Later on, it becomes a little more cognitively expensive or may even feel a little embarrassing not to know things. (It's not, though! Learning is great!) Don't waste the brief period where it's easy to not know.

Use your strengths

Remember how, when we talked about strategy in Chapter 3, I said you should build a strategy around your advantages? That's true here, too. You're going to want to pour a lot of information into your brain as efficiently as possible, so use your core muscles. If you're most comfortable with code, jump in. If you tend to go first to relationships, talk to people. If you're a reader, go get the documents.

Probably your preferred place to start won't give you all of the information you need, but it'll be a good place to start convincing your brain that this is just another project. Seriously, you've got this.

BUILDING CONTEXT

The start of a project will be full of ambiguity. You can create perspective, for yourself and others, by taking on a mapping exercise like we did in Chapter 2. That means building your *locator map*: putting the work in perspective; understanding the goals, constraints, and history of the project; and being clear about how it ties back to business goals. It means filling out your *topographical map*: identifying the terrain you're crossing and the local politics there, how the people on the project like to work, and how decisions will get made. And of course you'll need a *treasure map* that shows where you're all going and what milestones you'll be stopping at along the way.

Here are some points of context you'll need to clarify for yourself and for everyone else:

Goals

Why are you doing this project? Out of all of the possible business goals, the technical investments, and the pending tasks, why is *this* the one that's happening? The "why" is going to be a motivator and a guide throughout the project. If you're setting off to do something and you don't know *why*, chances are you'll do the wrong thing. You might complete the work without solving the real problem you were intended to solve. I'll talk about this phenomenon more in Chapter 6.

Understanding the "why" might even make you reject the premise of the project: if the project you've been asked to lead won't actually achieve the goal, completing it would be a waste of everyone's time. Better to find out early.

Customer needs

A story I tell a lot is about my first week in a new infrastructure team. A member of the team described a project they were working on, upgrading some system to make a new feature available. Another team, he said, needed the feature. "Why do they need it?" I asked, glad of an opportunity to get the lay of the land. "Maybe they don't," he said. "We think they do, but we have no way of knowing." These two teams sat in the same building, *on the same floor*.

Even on the most internal project, you have "customers": someone is going to use the things you're creating. Sometimes you'll be your own customer. Most of the time it's going to be other people. If you don't understand what your customers need, you're not going to build the right thing. And if you don't have a product manager, you're probably on the hook for figuring out what those needs are. That means talking to your customers and listening to what they say in response.

Product management is a huge and difficult discipline, and it's not easy to understand what your users actually want—as opposed to what they're telling you.[4] It takes time, so budget that time. Ask a user to let you shadow them using the software you're replacing. Ask internal users to describe the API they wish they had, or show them a sketch of the interface you think they want and see how they interact with it. Don't mentally fill in what you wish they'd said; listen to their actual responses. Try not to use jargon, because people can get intimidated and not want to tell you that they didn't understand. If you're lucky enough to have UX researchers on your team to study the customer experience, make sure to read their work, talk to them, and try to observe some user interviews.

Even if you *do* have a product manager, that doesn't mean you get to ignore your customers! I love the conversations with product managers that Gergely Orosz, author of the newsletter The Pragmatic Engineer, aggregates in his article "Working with Product Managers: Advice from PMs" (*https://oreil.ly/l9ofc*), especially Ebi Atawodi's comment that "You are also 'product.'" Atawodi points out that engineering teams should be just as customer-obsessed as product teams, caring about business context, key metrics, and the customer experience.

Success metrics

Describe how you'll measure your success. If you're creating a new feature, maybe there's already a proposed way to measure success, like a product requirements document (*https://oreil.ly/YBaZO*) (PRD). If not, you might be proposing your own metrics. Either way, you'll need to make sure that your sponsor and any other leads on the project agree on them.

Success metrics aren't always obvious. Software projects sometimes implicitly measure progress by how much of the code is written, but the existence of code tells you nothing about whether any problem has actually been solved. In some cases, the real success will come from *deleting* lines of code. Think about

---

4 Although it's almost certainly apocryphal, there's a Henry Ford quote about the Model T that illustrates how people in different domains can fail to communicate: "If I had asked people what they wanted, they would have said faster horses."

what success will really look like for your project. Will it mean more revenue from users, fewer outages, a process that takes less time? Is there an objective metric you can set up now that will let you compare before and after? In her Kubecon keynote, "The Challenges of Migrating 150+ Microservices to Kubernetes" (*https://oreil.ly/OTILj*), microservices expert Sarah Wells spoke about judging the success of a migration in two measurable ways: the amount of time spent keeping the cluster healthy, and the number of snarky messages from team members on Slack about functionality that didn't work as expected.

If you initiated the project, be even more disciplined about defining success metrics. If your credibility and social capital are strong, you can sometimes convince other people to get behind a project based on their belief in you or by means of a compelling document or an inspirational speech. But you can't be certain that you're right! Treat your own ideas with the most skepticism and get real, measurable goals in place quickly, so you can see how the project is trending. As Chapter 4 explained, your credibility and social capital can go down as well as up. Don't rely on them as the only motivator to keep the project going.

Sponsors, stakeholders, and customers

Who wants this project and who's paying for it? Who are the main customers of the project? Are they internal or external? What do they want? Is there an intermediate person between you and the original project sponsor? If there's a product requirements document, this may all be spelled out, but you might have to clarify for yourself who your first customer or main stakeholder is, what they're hoping to see from you, and when. If the impetus for the work has come from you, then you might be on the hook to continually justify the project and make sure it stays funded. It will be easier to sell the value of the work if you can find other people who want it too.

Fixed constraints

Maybe there are some senior technical roles where you can walk in and start solving big problems, unconstrained by budget, time, difficult people, or other annoying aspects of reality. I've never seen a role like that, though. Usually you're going to be constrained in some ways: understand what those constraints are. Are there deadlines that absolutely can't move? Do you have a budget? Are there teams you depend on that might be too busy to help you, or system components you won't be able to use? Will you have to work with difficult people?

Understanding your constraints will set your own expectations and other people's too. There's a big difference between "ship a feature" and "ship a feature without enough engineers and with two stakeholders who disagree on the direction." Similarly, creating an internal platform for teams that are eager to beta test it is a different project from trying to convince a hundred engineers to migrate to a new system that they hate. Describe the reality of the situation you're in, so you won't spend all your time being mad at reality for not being as you wish it to be.[5]

Risks

Is this a "moonshot" or a "roofshot" project? Does it feel huge and aspirational, or a fairly straightforward step in the right direction? In an ideal world, everyone on the project would deliver their own part in perfect synchronization, with predictable availability of time and energy (and ideally a boost to credibility, skills, quality of life, and social capital along the way!). The reality is that some things *will* go wrong, and the more ambitious the project, the riskier it will be. Try to predict some of the risks. What could happen that could prevent you from reaching your goals on deadline? Are there unknowns, dependencies, key people who will doom the project if they quit? You can mitigate risk by being clear about your areas of uncertainty. What don't you know, and what approaches can you take that will make those less ambiguous? Is this the sort of thing you can prototype? Has someone done this kind of project before?

One of the most common risks is the fear of wasted effort, of creating something that ends up never getting used. If you make frequent iterative changes, you have a better chance of getting user feedback and course-correcting (or even canceling the project early; we'll look at that more in Chapter 6) than if you have a single win-or-lose release at the end.

History

Even if this is a brand-new project, there's going to be some historical context you need to know. Where did the idea for the project come from? Has it been announced in an all-hands meeting or email that has set expectations? If the project is not brand new, its history may be murky and fraught. When teams have already tried and failed to solve a problem, there might be leftover components you'll be expected to use or build on, or existing users with odd use cases who will want you to keep supporting them. You might also face resentment and

---

5 It probably goes without saying, but be diplomatic when you describe this reality. If you write down the most charitable description of the difficult person, antagonistic team, or indecisive director, you'll feel less awkward when someone inevitably forwards the email or document to them. You'll build empathy for them too and perhaps have some insights about how to work with them.

irritation from the people who tried and failed, and you'll need to proceed very carefully if you want to engage their enthusiasm again.

If you're new to an existing project, don't just jump in. Have a lot of conversations. Find out what half-built systems you're going to have to use, work around, or clean up before you can start creating a new solution. Understand people's feelings and expectations, and learn from their experiences. Remember that tenet of Amazon's principal engineer community I mentioned in Chapter 2: "Respect what came before."

Team

Depending on the size of the project, you might have a few key people to get to know or a massive cast of team members, leads, stakeholders, customers, and people in nearby roles, some of whom influence your direction, some who'll make decisions you have to react to, and some of whom you'll never speak with directly. There'll also be other people in leadership roles.

If you're the lead of a project that only includes one team, you'll probably talk regularly with everyone on that team. On a bigger project with many teams involved, you'll need a contact person on each team. For even bigger projects, you might have a sublead in each area (see Figure 5-1). Or maybe your project is just one part of a broader project and *you'll* be a sublead.

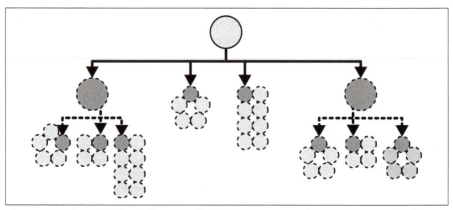

Figure 5-1. As the project lead (the root node of this tree) you may have contacts or subleads on each of several other teams. Some of those subleads may be directing work for their own subleads.

It's vital to build good working relationships with all of the other leaders and help each other out. Don't waste your time in power struggles. You'll be more likely to achieve your shared goals if you work well together, and of course work is a much more pleasant endeavor (higher quality of life, higher energy!) when you're harmonious with the people around you. Unfortunately, having multiple leaders often means unclear expectations about who's doing what—a common source of conflict. Understand who the leaders are, how they're involved in the project, and what role they expect to play.

GIVING YOUR PROJECT STRUCTURE

With all of that context in mind, you can start setting up the formal structures that will help you run the project. Setting expectations and structure and following a plan can be time-consuming, but they really do increase the likelihood that the thing you're eager to start working on will actually succeed. The more people involved, the more you'll want to make sure you're all aligned on your expectations. These structures will also act as tools to help you feel in control of what's going on—so if you're still a little overwhelmed, don't worry, this will make it easier.

Here are some of the things you'll do to set up a project.

Defining roles

I mentioned the risk of conflict when there are multiple leaders, so let's start there. At senior levels, engineering roles start to blur into each other: the difference between, say, a very senior engineer, an engineering manager, and a technical program manager might not be immediately clear. At a baseline, all of them have some responsibility to be the "grown-up in the room," identify risks, remove blockers, and solve problems. By definition, the manager has direct reports, but the engineer might too. The program manager sees the gaps, communicates about the project status, and removes blockers, but everyone else should step up and do those things if they aren't happening. We might argue that the engineer should have deeper technical skills, but some program managers are in deeply technical roles and many have extensive software engineering experience. This is even more complicated when a manager or TPM comes from an engineering background and is still involved in big technical decisions, or when there's more than one staff engineer. Who does what?

The beginning of a project is the best time to lay out each leader's responsibilities. Rather than waiting until two people discover they're doing the same job, or work is slipping through the cracks because nobody thinks it's theirs, you can describe what kinds of things will need to be done and who will do them. The simplest approach is to create a table of leadership responsibilities and lay out who should take on each one. Table 5-1 gives an example.

Table 5-1. Example table of leadership responsibilities

Product Manager	Olayemi
Technical Lead	Jaya
Engineering Manager	Kai
Technical Program Manager	Nana
Engineering Team	Adel, Sam, Kravann
Understanding customer needs and providing initial requirements	Product Manager
Providing KPIs for product success	Product Manager
Setting timelines	Technical Program Manager
Setting scope and milestones	Product Manager, Engineering Manager
Recruiting new team members	Engineering Manager
Monitoring and ensuring team health	Engineering Manager
Managing team members' performance and growth	Engineering Manager
Mentoring and coaching on technical topics	Technical Lead
Designing high-level architecture	Technical Lead (with support from engineering team)
Designing individual components	Technical Lead, Engineering Team
Coding	Engineering Team (with support from Technical Lead)
Testing	Engineering Team (with support from Technical Lead)
Operating, deploying, and monitoring systems	Engineering Team, Technical Lead
Communicating status to stakeholders	Technical Program Manager
Devising A/B experiments	Product Manager
Making final decisions on technical approach	Technical Lead
Making final decisions on user-visible behavior	Product Manager

I really want to emphasize that this is just an example! Some projects will have many more leaders than this. Some will have fewer. Internal-facing projects, like on infrastructure teams, usually won't have product managers.[6] If you would have put different names on different tasks here, that's *fine*. If you want to make all of this more sophisticated, a popular tool is RACI, also known as a responsibility assignment matrix (*https://oreil.ly/eebGs*). Its name comes from the four key responsibilities most typically used:

Responsible
> The person actually doing the work.

Accountable
> The person ultimately delivering the work and responsible for signing off that it's done. There's supposed to be only one accountable person per task, and it will often be the same person as whoever is "Responsible."

Consulted
> People who are asked for their opinion.

Informed
> People who will be kept up to date on progress.

If you're really into project management, you'll probably enjoy reading about the many, many variants of RACI, but I'm not going into that here. I'll just note that RACI turns the preceding list into a matrix, so you can set everyone's expectations even more clearly. It can be overkill for some situations, but when you need it, you *really* need it. A staff engineer friend at Google told me about using RACI for a chaotic project:

> We needed some kind of formal framework to explicitly define who would be making decisions. This helped us break out of two bad patterns: never making a decision because we didn't know who the decider was, so we'd just discuss forever, and relitigating every decision over and over because we didn't have a process for making decisions. RACI didn't solve either of those problems entirely, but it at least provided some (fairly uncontroversial) structure for people.

The uncontroversial structure is the real superpower here. It gives you a way to broach the conversation without it being weird.

---

6 Though think how much better our industry's internal solutions would be if they did.

Lara Hogan offers an alternative tool for product engineering projects, the team leader Venn diagram (*https://oreil.ly/Eoi2I*), which has overlapping circles for the stories of "what," "how," and "why," which are then assigned to the engineering manager, the engineering lead, and the product manager. I've heard suggestions that a fourth circle, the story of "when," if you could squeeze it into that diagram, might be assigned to a project or program manager.

However you approach it, try to get every leader aligned on what your roles are and who's doing what. If you aren't sure that everyone knows you're the lead (or even that you are), then the stress of a new project gets even worse. As a sub-lead for a project more than a decade ago, I found myself showing up at my weekly meetings with the overall lead a little nervous about whether I'd done what was expected of me. I could have saved myself a whole lot of anxiety by having a direct conversation about it: "Here's what I think I'm responsible for. Do you agree? Am I taking the right amount of ownership?"

Last thought on roles: if you're the project lead, you are ultimately responsible for the project. That means you're implicitly filling any roles that don't already have someone in them, or at least making sure the work gets done. If your teammates have no manager, you're going to be helping them grow. If there's nobody tracking user requirements, that's you. If nobody is project managing, that's you as well. It can add up to a lot. For the rest of this section, I'm going to talk about some tasks that may be assigned to you in these roles.

Recruiting people

If there are unfilled roles that you don't want to do or don't have time to do, you may have to find someone else to do them. That might mean recruiting someone internally or externally, or picking subleads to be responsible for parts of your project. Sometimes that will mean you're looking for specific technical skills that your team doesn't have enough of or experience you don't have.[7] Look also for the people who complement or fill the gaps in your own skill set. If you're a big-picture person, look for someone who loves getting into the details, and vice versa. For bonus points, see if you can find people who absolutely *love* doing the kind of work that you hate to do. That's the best kind of partnership!

I had an opportunity to lead a panel for LeadDev in October 2020, "Sustaining and Growing Motivation Across Projects" (*https://oreil.ly/o6Rj1*). In it, Mohit

---

7 Though try to avoid relying on one person's very specific skill set. People move between companies often, and you don't want a single point of failure.

Cheppudira, principal engineer at Google, talked about what he looks for when he recruits people:

> When you're responsible for a really big project, you're kind of building an organization and you're steering an organization. It's important to get the needs of the organization right. I spent a lot of time trying to make sure that I had the best leads in all the different domains that were involved in that one specific project. And, when you're looking for leads, you're looking for people that don't just have good technical judgment, but also have the attitudes: they are optimistic, good at conflict resolution, good at communication. You want people that you can rely on to actually drive this project forward.

Recruiting decisions are some of the most important you will make. The people you bring onto the project will make a huge difference in whether you meet your deadlines, complete visible tasks, and achieve your goals. Their success is your success, and their failure is very much your failure. Recruit people who will work together, push through friction, and get the job done—people you can rely on.

Agreeing on scope

Project managers sometimes use a model called the project management triangle (*https://oreil.ly/4aboG*), which balances a project's time, budget, and scope. You'll sometimes also hear this framed as "Fast, cheap, good: Pick two." It feels obvious to say, but it's somehow easy to forget: if you have fewer people, you can't do as much. Agree on what you're going to try to do.

Probably you're not going to deliver the whole project in one chunk. If you have multiple use cases or features, you'll want to deliver incremental value along the way. So decide what you're doing first, set a milestone, and put a date beside it. Describe what that milestone looks like: what features are included? What can a user do?

Jackie Benowitz, an engineering manager who has led several huge cross-organization projects, told me that she thinks about milestones as beta tests: every milestone is usable or demonstrable in some way, and gives the users or stakeholders an extra opportunity to give feedback. That means you have to be prepared for each incremental change to potentially change the user requirements for the next one, because changing what your users *can* do will help them realize what else they *want* to do. They might also tell you that you're on the wrong track, giving you an early opportunity to change your direction.

To maintain this kind of flexibility, some projects won't plan much further than the next milestone, considering each one to be a destination in itself. Others will roughly map out the entire project, updating the map when a change of direction is needed. Whichever you prefer, make the increments small enough that there's always a milestone in sight: it's motivational to have a goal that feels reachable. I've also found again and again that people don't act with a sense of urgency until there's a deadline that they can't avoid thinking about. Regular deliverables will discourage people from leaving everything until the end. Set clear expectations about what you expect to happen when.

If the project is big enough, you might split the work into *workstreams*, chunks of functionality that can be created in parallel (perhaps with different subteams), each with its own set of milestones. They may depend on each other at key junctures, and you may have streams that can't start until others are completely finished, but usually you can talk about any one of them independently from the others. You might also describe different *phases*, where you complete a huge piece of work, reorient, and then kick off the next stage of the project. Splitting the work up like this makes it a little more manageable to think about. It lets you add an abstraction and think at a higher altitude. If you can say a particular workstream is on track, for example, then you don't need to get into the weeds of each task on that stream.

If your company is using product management or roadmapping software, it will probably have functionality to organize your project into phases, workstreams, or milestones. If everyone who needs to participate is in one physical place, you can do the same thing with sticky notes on a whiteboard. What matters is that you all get the same clear picture of what you've decided you're doing and when.

Estimating time

I have met almost nobody who is good at time estimation. This may be the nature of software engineering: every project is different, and the only way we can tell how long a project will take is if we've done exactly the same thing before. The most common advice I've read is to break the work up into the smallest tasks you can, since those are easiest to estimate. The second most common is to assume you're wrong and multiply everything by three. Neither approach is very satisfying!

I prefer the advice Andy Hunt and Dave Thomas give in *The Pragmatic Programmer* (O'Reilly): "We find that often the only way to determine the timetable for a project is by gaining experience on that same project." As you deliver small slices of functionality, they explain, you gain experience in how long it will take your team to do something—so you update your schedules every time. They also recommend that you practice estimating and keep a log of how that's going. Like every other skill, the more you do it, the better you'll get at it, so practice estimating even when it doesn't matter and see if more of your estimates are right over time.

Estimating time needs to include thinking about teams you depend on. Some of these teams might be fully invested in the project, perhaps considering it their main priority for the quarter or year. Others may see it as just one of many requests competing for their attention. Talk with the teams you'll need as early as possible, and understand their availability.

Engineers in platform teams in particular have told me about the frustration of receiving a last-minute request to add functionality that's needed immediately for launch, functionality they could have easily provided if they'd known about it a few months earlier, when they could have incorporated it into their planning for the quarter. The later you tell other teams you'll need something from them, the less likely you are to get what you need. If they do agree to scramble to accommodate you, bear in mind that you're interrupting their previous work: you're disrupting the time estimation for their other projects.

Agreeing on logistics

There are a lot of small decisions that can help set your project up to run smoothly, and you'll probably want to discuss them as a team. Here are some examples:

When, where, and how you'll meet
> How often are you going to have meetings? If you're a single team, will you have daily standups? If you're working across multiple teams, how often will the leads get together? Will you have regular demos, agile ceremonies, retrospectives, or some other way to reflect?

How you'll encourage informal communication
> Meetings can be a fairly formal way to exchange information, and they probably don't happen every day. How can you make it easy for people to chat with each other and ask questions in the meantime? If you all sit

together, this can be pretty easy, but in an increasingly remote workforce, that's starting to become unusual.[8] In a new project where people don't know each other, they may be hesitant to send DMs, so you can encourage conversation with a social channel, informal icebreaker meetings, or (if people's lives allow it) getting everyone in the same place for a couple of days. Even silly things like meme threads can give people a connection that will make them quicker to ask each other questions and offer help.

How you'll all share status

How would your sponsor like to find out what's going on with the project? What about the rest of the company: where do you want them to go to find out more? If you're planning to send regular update emails, who will send them and at what cadence?

Where the documentation home will be

Does the project have an official home on the company wiki or documentation platform? If not, create one and make it easy to find with a memorable URL or prominent link. This documentation space will be the center of the project's universe and should link out to everything else. It will give you a single place to start when you're looking for a meeting recap, a description of the next milestone, or the wording of a related OKR. You want everyone to be looking at the same up-to-date information. It's a single fixed point in a chaotic universe!

What your development practices will be

In what languages are you going to work? How are you going to deploy whatever you create? What are your standards for code review? How tested should everything be? Are you releasing behind feature flags? If you're adding a new project in a company that has been around for a while, maybe there are standard answers for all of these questions. Others may depend on technical decisions you'll make as you work through the project. Begin the discussions, though, and get everyone aligned.

---

8 It's likely there are at least some people who are outside the office, and not uncommon for everyone on the team to be in a different city or time zone! If everyone is distributed around the globe, it's best if they have a good amount of overlap in their work days, especially if they will need to work together closely. It's hard to build a trusting relationship on completely asynchronous communication.

Having a kickoff meeting

The last thing you might do as part of setting up a project is to have a kickoff meeting. If all of the important information is written down already, this might feel unnecessary, but there's something about seeing each other's faces that starts a project with momentum. It gives everyone an opportunity to sync up and feel like part of a team.

Here are some topics you might cover at your kickoff:

- Who everyone is
- What the goals of the project are
- What's happened so far
- What structures you've all set up
- What's happening next
- What you want people to do[9]
- How people can ask questions and find out more

Driving the Project

My favorite talk about managing projects is "Avoid the Lake!" (*https://oreil.ly/ NHWFj*), by Kripa Krishnan, VP of Google Cloud Platform. I'd often heard the term "driving a project" without really thinking about what that means, but Krishnan makes the analogy clear when she says, "Driving doesn't mean you put your foot on the gas and you just go straight." Driving, in other words, can't be passive: it's an active, deliberate, mindful role. It means choosing your route, making decisions, and reacting to hazards on the road ahead. If you're the project lead, you're in the driver's seat. You're responsible for getting everyone safely to the destination.

In Chapter 3 we looked at one of the responsibilities of a project lead: making sure decisions get made. We're going to look now at some of the other challenges you might encounter on the road as you drive your project toward its destination.

---

9 Be clear about this point throughout the project! It's common for new leaders to sort of hint that it might be nice if everyone did something you need them to do. Think of it this way: if you're ambiguous, you're making *more* work for everyone else as they try to figure out how much the thing you've asked for matters. Be explicit about what you want people to do.

EXPLORING

I'm always suspicious when a brand-new project already has a design document or plan, and even more when those include implementation details: "Build a GraphQL server with Node.js to..." and so forth. Unless the problem is really straightforward (in which case, are you sure it needs a staff engineer?), you won't have enough information about it on day one to make these kinds of granular decisions. It will take some research and exploration to understand the project's true needs and evaluate the approaches you might take to achieve them. If you're creating a design where it's difficult to articulate the goals (or if the goals are just a description of your implementation!), that's a sign that you haven't spent enough time in this exploration stage.

What are the important aspects of the project?

What are you all setting out to achieve? The bigger the project, the more likely it is that different teams have different mental models of what you're trying to achieve, what will be different once you've achieved it, and what approach you're all taking. Some teams might have constraints that you don't know about, or unspoken assumptions about the direction the project will take: they might have only agreed to help you because they think your project will also achieve some other goal they care about—and they might be wrong! Team members may fixate on smaller, less important aspects of the project or niche use cases, or expect a different scope than you do. They may be using different vocabulary to describe the same thing, or using the same words but meaning something different. Get to the point where you can concisely explain what different teams in the project want in a way that they'll agree is accurate.

Aligning and framing the problem can take time and effort. It will involve talking to your users and stakeholders—and actually listening to what they say and exactly what words they use. It may involve researching other teams' work to understand if they're doing the same thing as you, just described differently. If you're going into this project with well-formed mental models, it can be difficult to set those preconceptions aside and explore how other people think about the work. But it will be even more painful to try to drive a project where everyone's using different words, or is aiming for a different destination.

As you explore, and uncover expectations, you'll start building up a crisp definition of what you're doing. Exploring helps you form an elevator pitch about the project, a way to sum it up and reduce it to its most important aspects. You'll also start building up a clear description of what you're not doing. Where projects are related to yours, you'll begin to show how one is a subset of the other, or how they overlap. It's clarifying to describe work that seems similar but isn't actually related, or work that seems entirely unrelated but has unexpected connections. I'll talk a little later in this chapter about building mental models to help you and others think about a problem in the same way. The better you understand the problem, the easier it will be to frame it for other people.

What possible approaches can you take?

Once you have a clear story for what you're trying to do, only then figure out how to do it. If you've gone into the project with an architecture or a solution in mind, it can be jarring to realize that it might not actually solve the real problem that you've framed as part of your exploration. This is such a difficult mental adjustment that I've seen project leads cling tightly to their original ideas about what problem they're solving, resisting all information that contradicts that worldview. That doesn't make for a good solution. So really try to keep an open mind about how you're solving the problem until you have agreed on what you need to solve.

Be open to existing solutions too, even if they're less interesting or convenient than creating something new. In Chapter 2, I talked about building perspective by studying and learning from other teams (both in your company and outside) before diving into creating some new thing. The existing work might not be exactly the shape of whatever you've been envisioning, but be receptive to the idea that it might be a better shape, or at least a workable one. Learn from history: understand whether similar projects have succeeded or failed, and where they struggled. Remember that creating code is just one of the stages of software engineering: running code needs to be maintained, operated, deployed, monitored, and someday deleted. If there's a solution that means your organization has fewer things to maintain after your project, weigh that up when you're choosing your approach.

CLARIFYING

A big part of starting the project will be giving everyone mental models for what you're all doing. When there are a lot of moving parts, opinions, and semirelated projects, it's a strain to keep track of them all. As the project lead, you have an incentive to spend time understanding the tricky concepts if it helps you achieve your project. But the people you ask for help have a different focus and may not try as hard. Unless you take the time to reduce the complexity for them, they could end up thinking about the project in a way that leads them to optimize for the wrong outcome or muddy a clear story you're trying to tell your organization.

In *The Art of Travel (Vintage)*, Alain de Botton talks about the frustration of learning new information that doesn't connect to anything you already know—like the sorts of facts you might pick up while visiting a historic building in a foreign land. He writes about visiting Madrid's Iglesia de San Francisco el Grande and learning that "the sixteenth-century stalls in the sacristy and chapter house come from the Cartuja de El Paular, the Carthusian monastery near Segovia." Without a connection back to something he was already familiar with, the description couldn't spark his excitement or curiosity. The new facts, he wrote, were "as useless and fugitive as necklace beads without a connecting chain."

I love that quote and think about it a lot while trying to help other people understand something. How can I hook this concept onto their existing knowledge? How can I make it relevant and spark their curiosity about it? Maybe I can build a necklace chain back via connecting concepts, or use an analogy to give them an idea that's close enough to be useful, even if it's not exactly correct.

Let's look at a few ways you can reduce the complexity of big messy projects by building shared understanding.

Mental models

When you start learning about Kubernetes, you're deluged with new terms: Pods, Services, Namespaces, Deployments, Kubelets, ReplicaSets, Controllers, Jobs, and so forth. Most documentation explains each of these concepts through its relationship to *other* new terms, or describes them in abstract ways that make perfect sense *if* you already understand the whole domain. If you're coming in cold, it can be overwhelming—until a friend frames it in relation to something familiar. They might use an analogy that lets you imagine the behavior of something you already know: "Think of this part like a UNIX process." They might use an example instead, to give you a hint to the shape of the concept being described: "This is likely to be a Docker container." Neither of these models is perfect, but they don't have to be: they have to be *close enough* to make a chain back to some other thing you already understand, to give you something to hook the knowledge onto.

I've deployed these sorts of rhetorical devices throughout this book, using video game analogies and geographical metaphors to describe concepts. Connecting an abstract idea back to something I understand well removes some of the cognitive cost of retaining and describing the idea. It's like I'm putting the idea into a well-named function that I can call again later without needing to think about its internals. (See, I just did it again.)

Just like we build APIs and interfaces to let us work with components without having to deal with their messy details, we can build abstractions to let us work with ideas. "Leader election" is something we can understand and explain more easily than "distributed consensus algorithm." As you describe the project you want to complete, you'll likely have a bunch of abstract concepts that aren't easy to understand without a whole lot of knowledge in the domain you're working in. Give people a head start by providing a convenient, memorable name for the concept, using an analogy, or connecting it back to something they already understand. That way they'll be able to quickly build their own mental models of what you're talking about.

Naming

Two people can use the same words and mean quite different things. I joke that conversations with one of my favorite colleagues always devolve into us arguing about the meanings of words. But once we understand each other, we can speak in a very nuanced, high-bandwidth way and have a much more powerful conversation about where we actually agree or disagree.

In 2003, Eric Evans wrote *Domain-Driven Design* (Addison Wesley) and gave us the concept of deliberately building what he called a "ubiquitous language" (*https://oreil.ly/CEQmR*): a language shared by the developers of a system and the real-world domain experts who are its stakeholders. Inside a company, even very common words like *user, customer,* and *account* may have specific meanings, and those can even change depending on whether you're talking to someone in finance, marketing, or engineering. Take the time to understand what words are meaningful to the people you intend to communicate with, and use their words when you can. If you're trying to talk with multiple groups at once, provide a glossary, or at least be deliberate about describing what *you* mean by the terms you're using.

Pictures and graphs

If you really want to reduce complexity, use pictures. There's no easier way to help people visualize what you're talking about. If something's changing, a set of "before" and "after" pictures can be clearer than an entire essay. If one idea fits within another, you can draw them as nested; if they're parallel concepts, they can be parallel shapes. If there's a hierarchy, you might depict it as a ladder, a tree, or a pyramid. If you're representing a human, using a stick figure or smiley-face emoji is clearer than just drawing a box.

Be aware of existing associations: don't use a cylinder on your diagram unless you're OK with many readers thinking of it as a datastore. If you use colors, some of your audience will try to interpret their meaning, for example assuming that green components are intended to be encouraged and red ones should be stopped.

Pictures can also take the form of graphs or charts. If you can show a goal and a line trending toward that goal (like in Figure 5-2), it's easy to see what success will look like. Similarly, if your line is trending toward some disaster point you're highlighting, the need for the project can become viscerally clear.

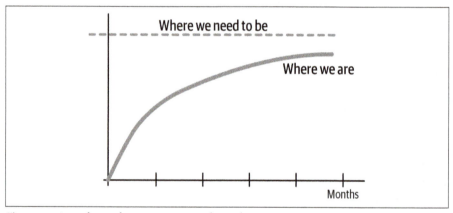

Figure 5-2. A graph can show progress toward a goal.

DESIGNING

Once the exploration is done and the work is clarified, you'll probably have a lot of ideas for what happens next: what you're going to build or change, and what approach you're going to take. Don't assume everyone you work with understands or agrees with those ideas. Even if you're not hearing objections when you talk about them, your colleagues may not have internalized the plan and their implicit agreement may not mean anything. You'll need to work to make sure everyone is aligned. The most efficient way to do that is to write things down.

Why share designs?

In Chapter 2 I discussed oral versus written company cultures. The bigger the company, the more likely that your culture has shifted toward the latter, and that there is some expectation that you will write and review design documentation.

That's because it's very difficult to have many people achieve something together without shared understanding, and it's hard to be sure you have that shared understanding without a written plan. Whether you're creating features, product plans, APIs, architecture, processes, configuration, or really anything else where multiple people need to have the same understanding, you won't truly know if people understand and agree until you write it down.

Writing it down doesn't mean you need a 20-page technology deep dive for every tiny change. A short, snappy, easy read can be perfect for getting a group on (literally) the same page. But you should at least include the important aspects of the plan, and let other people get in touch with you if they see hazards in your path. Asking for review on a design doesn't just mean asking about the feasibility of an architecture or a series of steps; it includes agreeing on whether you're solving the right problem at all and whether your assumptions about other teams and existing systems are correct. An idea that seems like an obvious path for one team may cause work for or break the workflows of another org. As my friend Cian Synnott says, a written design is a very cheap iteration (*https://oreil.ly/gp5f2*).

RFC templates

A common approach to sharing information in this way is a design document, often called a request for comment document (or RFC). Although you'll find RFCs used at many companies, there's not really a consistent standard for what one should look like or how they get used. Different companies will have different levels of detail and formality, you'll share more or less broadly, comments may or may not be encouraged, and you might have an official approval step or meeting to discuss the design.

I'm not going to weigh in on which process is best—it really depends on your culture—but I'm a big fan of having templates for documents like this.[10] No matter how amazing we are as architects, there's a lot to remember when designing a complex system or process or change. And humans aren't great at paying attention to all of the things. As Atul Gawande, author of *The Checklist Manifesto: How to Get Things Right* (Picador), says:

> *We are not built for discipline. We are built for novelty and excitement, not for careful attention to detail. Discipline is something we have to work at.*

---

10 I wrote a blog post for the Squarespace engineering blog that includes a sample template: "The Power of 'Yes, if': Iterating on Our RFC Process" (*https://oreil.ly/Pdb3l*).

It somehow feels beneath us to use a checklist, an embarrassment. It runs counter to deeply held beliefs about how the truly great among us—those we aspire to be—handle situations of high stakes and complexity. The truly great are daring. They improvise. They do not have protocols and checklists. Maybe our idea of heroism needs updating.[11]

Gawande argues that using a checklist helps us talk to each other, avoid common mistakes, and make the right decisions *intentionally* instead of implicitly. A good RFC template helps you think through the decisions and reminds you of topics you might otherwise forget. Going through the exercise of creating this kind of document and answering some (perhaps uncomfortable) questions about your plan will help ensure you haven't missed some vital category of problem.

What goes in an RFC?

Your company may already have its own RFC template, and you should follow that if it exists. However, here are the headings that I put in every RFC, and that I think should be included at an absolute minimum.

Context I like documents to be anchored in space and time. When someone stumbles across this document in two years, the header should give them enough context to decide whether it's relevant to whatever they're searching for. It should have a title, the author's name, and at least one date; I like "created on" and "last updated on," but either of those is better than none. Include the status of the document: whether it's an early idea, open for detailed review, superceded by another document, being implemented, completed, on hold. I like a standard format for headers so that scanning an RFC quickly is really easy, but it's more important that the information is available than that it's standardized.

Goals The goals section should explain why you're doing this at all: it should show what problem you're trying to solve or what opportunity you're trying to take advantage of. If there's a product brief or product requirements document, this section could be a summary of that, with a link back. If the goal just suggests the question, "OK, but why are you doing *that*?" then you should go a step further and answer that question too. Provide enough information to let your

---

11 He goes on to show that checklists save lives. Few people reading this will be responsible for life-critical systems, but if you are, please have protocols and checklists!

readers know whether they think you're solving the right problem. If they disagree, that's great—you found out now and not after you built the wrong thing.

The goal shouldn't include implementation details. If you send me an RFC with a goal of "Create a serverless API to translate the sounds of chickens," I can absolutely believe that that's what you're trying to do, and I can review the RFC and try to appraise your design. But without knowing what actual problem you're trying to solve and for whom, I can't evaluate whether this is really the right approach. You've specified in the goal that you're setting out to make it serverless, so you've already made a major design decision without justifying it. The specific implementation should *serve* the goal; it should not *be* the goal. Leave the design decisions to the design section.

Design The design section lays out how you intend to achieve the goal. Make sure that you include enough information for your readers to evaluate whether your solution will work. Give your audience what they need to know. If you're writing for potential users or product managers, make sure you're clear about the functionality and interfaces you intend to give them. If you'll depend on systems or components, include how you'll want to use them, so readers can point out misunderstandings about their capabilities.

Your design section could be a couple of paragraphs or it could be 10 dense pages. It could be a narrative, a set of bullet points, a bunch of subsections with headers, or any other format that will clearly convey the information.

Depending on what you're trying to do, the design section could include:

- APIs
- Pseudocode or code snippets
- Architectural diagrams
- Data models
- Wireframes or screenshots
- Steps in a process
- Mental models of how components fit together
- Organizational charts
- Vendor costs
- Dependencies on other systems

What matters is that at the end, your readers should understand what you intend to do and should be able to tell you whether they think it will work.

Wrong Is Better Than Vague

I've often seen people be a little hand-wavy in writing this design section —or avoid committing to a plan at all—because they don't want people to argue with them about the details. But it's a better use of your time to be wrong or controversial than it is to be vague. If you're wrong, people will tell you and you'll learn something, and you can change direction if you need to. If you're trying out a controversial idea, you can find out early whether your colleagues will pull against your approach. Having disagreements about your design doesn't mean that you need to change course, but it gives you information you wouldn't have had otherwise.

Here are two tips to make your design more precise:

- Be clear about who or what is doing the action for every single verb. If you find yourself writing in the passive voice, like "The data will be encrypted in transit" or "The JSON payload will be unpacked," then you're obscuring information and making the reader guess. Instead, write with active verbs that have a subject who does the action: "The client will encrypt the data before it is transmitted" or "The Parse component will unpack the JSON payload."[12]

- Here's a tip that was a game-changer for me: It's fine to use a few extra words or even repeat yourself if it means avoiding ambiguity. As software engineer and writer Eva Parish recommends in her post "What I Think About When I Edit" (*https://oreil.ly/TExXb*):

  Instead of saying "this" or "that," you should add a noun to spell out exactly what you're referring to, even if you've just mentioned it.

---

12 Dr. Rebecca Johnson offers (*https://oreil.ly/zSRhh*) the best test I've ever read for accidental use of the passive voice: "If you can insert 'by zombies' after the verb, you have passive voice." It almost always works. "The data will be encrypted in transit [by zombies]." Thanks, zombies!

> *Example: We only have two boxes left. To solve this, we should order more.*
>
> *Revision: We only have two boxes left. To solve this shortage, we should order more.*

Since I read Eva's article, I notice so many examples where a bare "this" or "that" in a design document obscures information. For example, "A proposal exists to replace OldSolution, which was built to provide OriginalFunctionality, with NewSolution. TeamB needs this, so we should discuss requirements." What does TeamB need: the proposal, the original functionality, or the new solution?

If you struggle with writing, bear in mind that it's a learnable skill. Any learning platform your company has access to will probably have a technical writing class on offer. Or consider Google's courses, Technical Writing One and Technical Writing Two (*https://oreil.ly/DUTL2*). The Write the Docs (*https://oreil.ly/g535C*) website also offers a ton of resources on how to write well.

Security/privacy/compliance What do you have worth protecting, and from whom are you protecting it? Does your plan touch or collect user data in any way? Does it open up new access points to the outside world? How are you storing any keys or passwords you're going to use? Are you protecting against insider or external threats, or both? Even if you think there are no security concerns and believe this section isn't relevant, write down why you believe that is the case.

Alternatives considered/prior art If you could solve this same problem with spreadsheets, would you still want to do it?[13] The "alternatives considered" section is where you demonstrate (to yourself and others!) that you're here to solve the problem and you aren't just excited about the solution. If you find yourself omitting this section because you *didn't* consider any alternatives, that's a signal that you may not have thought the problem through. Why *wouldn't* simpler solutions or off-the-shelf products work? Has anyone else at your company ever tried something similar, and why isn't their solution a good fit? I have a policy that if a

---

13 If you love spreadsheets and this just makes the project more attractive to you, sub in whatever technology you find most mundane.

plausible-seeming option already exists inside the company and we're not going to use it, the RFC author *has to* send the new design to the people who own that system and give them an opportunity to respond.

Those are the headings that I think absolutely have to be included, even on a tiny RFC. They're the "keep you honest" sections! But there are some others that are often helpful if you want to get the most value out of the document you're writing.

Background What's going on here? What information does a reader need to evaluate this design? You could include a glossary if you're using internal project names, acronyms, or niche technology terms that reviewers might not know.

Trade-offs What are the disadvantages of your design? What trade-offs are you intentionally making because you think the downsides are worth the benefits?

Risks What could go wrong? What's the worst that could happen? If you're a bit nervous about system complexity, added latency, or the team's lack of experience with a technology, don't hide that concern: warn your reviewers and give them enough information to draw their own conclusions about it.

Dependencies What are you expecting from other teams? If you're going to need another team to provision infrastructure or write code, or if you need security, legal, or comms to approve your project, how much time will you need to allow them? Do they know you're coming?

Operations If you're writing a new system, who will run it? How will you monitor it? If it will need backups or disaster recovery tests, who will be responsible for those?

Technical pitfalls

While this is not intended to be a technical or architectural book, I do want to call out a few pitfalls I often see in design documentation. Catch them for yourself so other people don't have to.

It's a brand-new problem (but it isn't) There are occasional exceptions, but your problem is almost certainly not brand new. I already talked about looking for prior and related projects, but it's important enough to mention again here. Don't miss the opportunity to learn from other people, and consider reusing existing solutions.

This looks easy! Some projects are seductively harder than they look, and you might not realize that until you're deep in the weeds of implementing them. Software engineers don't always really internalize that other domains are as rich and nuanced and complex as their own. They might see, say, an accounting system and assume they can build a better, cleaner, simpler one. How could previous teams have put thousands of engineer-hours into *this*!? But building an accounting system (or a payroll system, or a recruitment system, or even something to correctly share on-call schedules) is actually a hard problem. If it seems trivial, it's because you don't understand it.

Building for the present If you're building for the state of the world as it is right now, will your solution still work in three years? Even if you're designing for five times the current number of users and requests, there are other dimensions to think about. If the system needs to know about all teams or all products at your company, what happens as the company grows, or has acquisitions, or is acquired? If you have five times as many products, will this component become a bottleneck as everyone waits for one team to add custom logic for them? If your team doubles in size, will its members still be able to work in this codebase?

Building for the distant, distant future If you're designing for a few orders of magnitude more than your current usage, do you have a real reason to go that big? If it's trivial to handle more users, that's great, do it, but watch out for over-engineered solutions that are much more complicated than they need to be. If you're adding custom load balancing, extra caching, or automatic region failover, explain why it's worth the extra time and effort. "We might need it later" is not a good enough justification.[14]

Every user just needs to… If you have five users, you can probably individually teach each of them all the arcane rules of your system. If you have hundreds, or more, they're going to do it wrong; if you don't plan for that, your design doesn't work. Any part of your solution that involves humans changing their workflows or behavior will be difficult and needs to be part of the design.

We'll figure out the difficult part later This one is common in migrations: you spend a quarter building and deploying the system, polishing it up, and making it perfect for a couple of easy use cases—and then you have to figure out how to make it work for more difficult cases. What happens if that turns out not to be

---

14 One of the most expensive phrases ever uttered in software engineering.

possible? Ignoring the difficult part of your project might also mean pushing complexity to someone else, like requiring every existing caller of an API to change their code rather than being backward compatible, or forcing your clients to write their own logic to interpret arcane and scattered information.

Solving the small problem by making the big problem more difficult If you have lots of tiny projects with barely enough staffing to scrape by, you'll see people working around difficult problems in hacky ways instead of engaging with them directly. These tacked-on solutions often have hidden dependencies on existing system behavior that mean it will be harder to implement a more comprehensive solution later. If your organization refuses to invest in solving the underlying problem, you may not have any choice, but at least call it out in your design. Think about how you can solve the smaller problem without making the bigger one less tractable.

It's not really a rewrite (but it is!) If you're looking at a huge software system and envisioning it in a different shape, be honest with yourself and others about how much work that will take. You might imagine "just" taking the business logic and refactoring it, for example, or rearchitecting it for the cloud. But unless your code is already very modular and well organized (in which case, are you sure you need to rearchitect?), chances are you'll end up rewriting a lot more than you intended. If your project is a veiled "rewrite from scratch," be honest with yourself and admit it.

But is it operable? If you struggle to remember how something works at 3 p.m., you won't understand it at 3 a.m. And the people who join your team after you've moved on will find it much harder. Make sure you create something that other people can reason about. Aim to make systems observable and debuggable. Make your processes as boring and as self-documenting as possible.

Speaking of operability, if it's going to run in production, decide who's on call for it and put that in the RFC. If that's your own team, make sure you have more than three people (ideally at least six) or you'll be setting yourselves up for burnout and dropped pages.

Discussing the smallest decisions the most Who doesn't love a good bikeshed discussion! The expression "bikeshedding" came from C. Northcote Parkinson's 1957 "Law of Triviality" (*https://oreil.ly/D8nvw*), which holds that since it's much easier to discuss a trivial issue than a difficult one, that's where teams tend to spend their time.[15] Parkinson's example was a fictional committee evaluating the plans for a nuclear power plant but spending the majority of its time on the easiest topics to grasp, like what materials to use for the staff bicycle shed.[16] Tech people are usually aware of the concept of bikeshedding, but even senior people drift into writing long paragraphs about the most trivial, reversible decisions, while not engaging at all with the ones that are harder to grasp or to find consensus on.

These are just a few of the common pitfalls. You've probably noticed others. Add those to your list and be really sure you're not falling prey to them yourself.

CODING

Most software projects will involve writing a lot of new code or changing existing code. In this section, I'll talk about how a project lead can engage with this kind of hands-on technical work. (If you're not a software engineer, swap in whatever core technical work makes sense for you here.)

Should you code on the project?

As the project lead, how much code you contribute will vary depending on the size of the project, the size of the team, and your own preferences. If you're on a tiny team, you might be deep in the weeds of every change. On a project with multiple teams, you might contribute occasional features, or just small fixes, or you might work at a higher level and not code at all. Many project leads find that they review a lot of code, but don't write much themselves.

As Joy Ebertz points out (*https://oreil.ly/mPHXC*), "Writing code is rarely the highest leverage thing you can spend your time on. Most of the code I write today could be written by someone much more junior." Ebertz notes, however, that coding gives you a depth of understanding that's hard to gain otherwise and helps you spot problems. What's more, "If spending a day a week coding keeps you engaged and excited to come to work, you will likely do better in the rest of your job." Finally, staying involved in the implementation ensures that you feel the cost of your own architectural decisions as much as your team does.

---

15 Parkinson also coined (*https://oreil.ly/AE5w5*) the law that "work expands so as to fill the time available for its completion." The dude was insightful!

16 I wrote a talk once where someone left no comments on the entire 83-slide deck except to suggest a replacement for the bikeshed picture I'd included on one of the slides. True story! I don't think they were being ironic.

Notice, though, if you're contributing code at the expense of more difficult, more important things. This is a form of the *snacking* I mentioned in Chapter 4: taking on work that you know how to do (and that has a shorter feedback loop) and avoiding the big, difficult design decisions or crucial organizational maneuvering.

Be an exemplar, but not a bottleneck

As the person responsible for moving the project along, your time is going to be less predictable than other people's. You'll probably have more meetings than everyone else does too. So if you take on the biggest, most important problems, chances are you'll take longer to get to coding work than someone else would, which can block others and make you a bottleneck. If you're coding, try to pick work that's not time-sensitive or on the critical path.

Think of your code as a lever to help everyone else. Katrina Owen, coauthor of *99 Bottles of OOP* (*https://oreil.ly/kEpkE*) and a staff engineer at GitHub, told me about a project where she created a standard way of writing a test for API pagination, then replaced all of the existing tests with her approach. By changing all of the current pagination tests, she was implicitly improving future tests too: anyone creating one would copy the pattern that was already there.

Aim for your solutions to empower your team, not to take over from them. Ross Donaldson, a staff engineer working on database systems, has described part of his work to me as "scouting and cartography":

> I come back to the team and say, "I found this problem, that river, these resources," then we can all together discuss how we want to approach this new information. Then maybe I go out and build a rough bridge over the new river, which the team will own and improve. I provide an opinion or two and remind people of some of the tools they have at their disposal, but otherwise prioritize their sense of ownership over my own sense of code aesthetic.

Polina Giralt, senior staff engineer at Squarespace, adds,

> If there's something only I understand, I'll do it, but insist that someone pairs up with me. Or if it's an emergency and I know how to fix it, I'll do it myself and explain it later. Or I'll write the code to establish a new pattern, but then hand it off to someone else to continue implementing it. That way it forces knowledge sharing.

Rather than taking on every important change yourself, find opportunities for other people to grow, by chatting over the details or pair programming on the change. Pairing shares knowledge and builds other people's skills. Pairing also means you can dip in for the key part of a change, then leave your colleague to complete the work.

If you're reviewing code and changes, be aware of how your comments are received. Even if you think you're relatable and friendly, it can be intimidating for early-career engineers when a staff engineer comments on their work. You want the rest of the team to think of you as a resource to learn from, not as someone who criticizes every decision and makes them feel inadequate. Sometimes it's better to let someone set a pattern that's *good enough* and not overrule them, even if you would have done it better. Also, be careful that you're not doing *all* the code reviews—or you'll be a single point of failure and the rest of your team won't learn as much.

A staff+ engineer is an exemplar in another, more implicit way: *whatever you do will set expectations for the team.* For that reason, it's important to produce good work. Meet or exceed your testing standards, add useful comments and documentation, and be very careful about taking shortcuts. If you're the most senior person on the team and you're sloppy, you're going to have a sloppy team. (I'll talk more about being a role model in Chapter 7.)

COMMUNICATING

Communicating well is key for delivering a project in time. If the teams aren't talking to each other, you won't make progress. And if you're not talking to people outside the project, your stakeholders won't know how it's going. Let's look at both of these types of communication.

Talking to each other

Find opportunities for your team members to talk with each other regularly and build relationships, even (or maybe especially) on fully remote teams. It should feel easy to reach out and ask questions, and they should be comfortable enough with each other that they can disagree without it getting tense. You can make relationship building easier with shared meetings, friendly Slack channels, demos, and social events. If you have a small number of teams, key people from adjacent teams can attend each other's meetings or standups. The goal is comfortable familiarity.

Familiarity will make it feel safer to ask clarifying questions too. Engineers who don't know each other may feel uncomfortable saying "I don't know what that term means" or "What are the implications of the problem you just described?" It will be harder to work together, share knowledge, and uncover misunderstandings as a result. Aim to get to a place where it's normal for your team to ask questions and admit what they don't know.

Sharing status

Your project has other people who care about it: stakeholders, sponsors, customer teams who are waiting for you to be done. Make it easy for them to find out what's going on, and set their expectations about when you'll reach various milestones. That might mean one-on-one conversations, regular group email updates, or a project dashboard with statuses.

As you understand the progress of your project, you'll probably pick up a lot of detail and nuance about who's doing what, what's intended to happen when, and how each part of the project is going. When you're delivering status updates, you might feel inclined to share everything you know: more information is better, right? Not necessarily! Too much detail can obscure the overall message and make it harder for your audience to take away the conclusion you intended.

Instead, explain your status in terms of impact and think about what the audience will actually *want* to know. They probably don't care that you stood up three microservices; they care about what users can do now, and when they'll be able to do the next thing. If it's becoming clear that you're not going to hit a key milestone date, that's a fact you'll want to pass on. But if one team is delayed in a way that doesn't change your delivery date, that delay may not be relevant to them. Calling it out may even look like you're trying to escalate something that doesn't need escalation.

If you think your audience really will want all the details, at least lead with the headlines. Don't assume that they'll sift through your update for the key facts or read between the lines to pick up nuances. If something is not clear, spell out the takeaways. Practice following facts with "That means..." or explaining that you're doing something "so that we can..."

Be realistic and honest about the status you're reporting. If your project is having difficulties, it may be tempting to put on a brave face, hope you'll sort everything out, and report that the status is green. When you do this, though, you risk an unpleasant surprise at the end of the project when you have to admit that it's not and hasn't been for a while. Have you ever heard someone talk about a "watermelon" project? They're all green on the outside, but the inside is red. If your project is stuck, don't hide it: ask for help.

NAVIGATING

Something will always go wrong. Maybe you realize that a technology that's been core to your plans isn't going to be a good fit after all: it won't scale, it's missing some table-stakes feature, or it's got a licensing condition your legal team has absolutely vetoed. Maybe your organization announces a change in business direction and now needs you to solve a different problem. Maybe someone vital to the project quits. It is inevitable that you'll meet some roadblocks and have to change direction, and you'll have a better time if you go into the project assuming *something* is going to go wrong—you just don't know yet what it will be. This attitude can help you be more flexible, so you'll find it sort of interesting when the roadblock arrives, rather than being frustrated by it. Reframe these diversions as an opportunity to learn and to have an experience you wouldn't have had otherwise.

As the person at the wheel, you're accountable for what happens when your project meets an obstacle. You don't get to say, "Well, the project is blocked and so there's nothing we can do": you are responsible for rerouting, escalating to someone who can help, or, if you really have to, breaking the news to your stakeholders that the goal is now unreachable. Avoid those "watermelon projects": if the project status is green apart from one key problem that's going to be impossible to solve, the project status is not really green!

Whatever the disruption, work with your team to figure out how you can navigate around it. The bigger the change, the more likely it is that you'll need to go back to the start of this chapter, deal with the overwhelm all over again, build context, and treat the project like it's restarting. Whatever happens, make sure you keep communicating well about it. Don't create panic or let rumors start. Instead, give people facts and be clear about what you want them to do with the information.

When you're having difficulties, remember that you are not the only person who wants your project to succeed.[17] Your manager's job is to make you

---

17 Unless you're really spending your social capital on a passion project that nobody else cares about, as discussed in Chapter 4! In that case, sorry, you're probably on your own. But I hope you've got enough goodwill that your colleagues and manager are still sort of enthusiastic about it on your behalf and are willing to help you get it unstuck.

successful, and your director's job is to make your organization successful. If you're not telling them you need help, it's going to be harder for them to do their jobs. Some people really resist asking for help. It feels like failure, maybe. But if you're stuck and need help, the biggest failure is not asking for it. Don't struggle alone.

I'll talk about navigating obstacles a lot more in the next chapter, when we look at some of the reasons that projects can get stuck and how to get them back on track.

To Recap

- Staff engineers can take on problems that seem intractable and make them tractable.

- It's normal to feel overwhelmed by a huge project. The project is difficult. That's why it needs someone like you on it.

- Set up the structures that will reduce ambiguity and make it easy to share context.

- Be clear on what success on the project will look like and how you'll measure it.

- Leading a project means deliberately *driving* it, not just letting things happen.

- Smooth your path by building relationships and deliberately setting out to build trust.

- Write things down. Be clear and opinionated. Wrong gets corrected, vague sticks around.

- There will always be trade-offs. Be clear what you're optimizing for when you make decisions.

- Communicate frequently with your audience in mind.

- Expect problems to arise. Make plans that assume there will be changes in direction, people quitting, and unavailable dependencies.

6

Why Have We Stopped?

As the project's driver, you're responsible for getting everyone safely to the destination. But there are a lot of reasons your journey might stop early. You could run into roadblocks: accidents, toll booths, or a country road full of sheep. You might find that you've lost your map, or that the various people in the car disagree about where you're going. Or you might just realize you should be going somewhere else.

The Project Isn't Moving—Should It Be?

In Chapter 5, we talked about starting projects. Now it's time to look at ways they can halt. We'll start with two kinds of temporary stops you might encounter when something's going wrong: getting blocked by something, and getting lost.

Then we'll look at ways you might intentionally stop the journey. Sometimes this means declaring victory too soon, but it's also sometimes just time for the journey to end, whether it's reached its destination or not.

This is a good opportunity to note that, as a leader in your organization, you can help projects that you're *not* leading too. Sometimes the best use of your time will be to set aside what you were doing and use pushes, nudges, and small steps (and, OK, sometimes major escalations) to get a stalled project moving again. As Will Larson says (*https://oreil.ly/LKcoI*), this small investment of time can have a huge impact:

> *A surprising number of projects are one small change away from succeeding, one quick modification away from unlocking a new opportunity, or one conversation away from consensus. With your organizational privilege, relationships you've built across the company, and ability to see*

around corners derived from your experience, you can often shift a project's outcomes by investing the smallest ounce of effort, and this is some of the most valuable work you can do.

For consistency, the rest of this chapter will assume that you're leading the project that has stopped. However, a lot of these techniques will work just as well if you're stepping in to help.

Don't forget, though: seeing a problem doesn't necessarily mean you should jump on it. As I said in Chapter 4, you need to *defend your time.* Don't get into the situation depicted in the time graph in Figure 6-1, where you're caught up in so many side quests and assists that you have no time for the work you're accountable for. Be discerning! Choose the opportunities where your help is most valuable, and then take deliberate action, with a plan for stepping away again afterward.

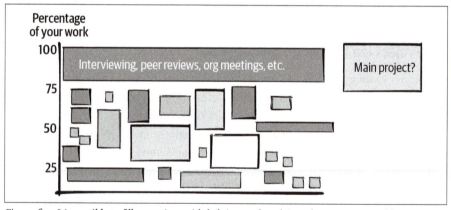

Figure 6-1. It's possible to fill your time with helping and nudging along projects and leave no space for the main project you promised to do.

Let's start with the first set of reasons projects stall: they get blocked.

YOU'RE STUCK IN TRAFFIC

In a perfect world, teams would be autonomous and never have to think about each other's work. In reality, any big project is going to span multiple teams, departments, and job functions. Even a single person's procrastination can sometimes be enough to miss an important deadline. And when the project is a migration or a deprecation, success might depend on work from *every* other engineering team, with many opportunities to get blocked along the way.

No matter what kind of blockage you're dealing with, you'll use some of the same techniques:

Understand and explain

You'll debug what's going on and understand the blockage. Then you'll make sure everyone else has the same understanding of what's happening.

Make the work easier

You'll work around blockages by not needing as much from the people you're waiting for.

Get organizational support

It's easier to get work prioritized when you can show that it's an organizational objective. You'll demonstrate the value of the work so that you can get that support. And sometimes you'll escalate to get help getting past a blockage.

Make alternative plans

Sometimes the blockage just isn't going to go away, and you'll use creative solutions to succeed. Or you'll accept that the project just can't happen in its current form.

Let's look at how to use these techniques when you're blocked in various ways: waiting for another team, a decision, an approval, a single person, a project that's not assigned, or all of the teams involved in a migration.

BLOCKED BY ANOTHER TEAM

We'll start with a classic dependency problem: your project is on track but you need work from another team, and it's *not happening*. If you're lucky, a leader of that team is telling you up front what's going on and when they'll be ready to go. If you're less lucky, the team has stopped replying to your emails and you're piecing together what's going on. Waiting on them is frustrating. They have all the information you have, and they know there's a launch date! Why don't they care? Go find out.

What's going on?

If a team you depend on is not delivering what you need, there's almost certainly a great reason why. Three likely reasons are misunderstandings, misadventures, or misaligned priorities.

Misunderstandings

Even in organizations with clear communication paths, information can get lost. One team thinks it's obvious that something needs to happen by a specific date. The other team has no idea that there's a deadline, or has taken away a different interpretation of what they've been asked to do.

Misadventure

Life happens. Someone quits, or gets sick, or needs to take abrupt leave. The team you depend on is understaffed, overloaded, or blocked by their own downstream dependencies. It might be impossible for them to meet the deadline, no matter how important it is.

Misalignment

Maybe the team has impressive velocity—just not on your project. Even if you're working on something vital, the other team may have an even higher priority. Take a look at Figure 6-2. Project C is Team 2's highest priority, and they're focused on it above any other work they have to do. But it's only Team 1's third-highest priority! They'll get to it if they have time.

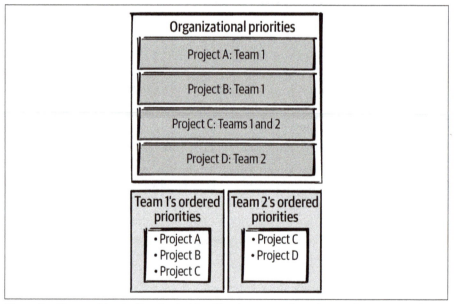

Figure 6-2. Teams with misaligned priorities. If each team works on its most important project, Team 2 will likely be waiting on Team 1.

Navigating the dependency

Here are those four techniques in action:

Understand and explain Start by understanding why the other team isn't moving. Do they not understand what's needed? Is something in their way? Understanding means talking to each other. If DMs or emails aren't working, take it to a synchronous voice conversation–yes, this means a meeting. If the team is hard to reach, going through back channels might help: hopefully you've built a bridge with someone on or near the team. (See Chapter 2.)

Explain why the work is important, and spell out what you're hoping they'll deliver and by when. Give them another chance to tell you whether that's realistic, or whether there's something else they can do that would solve your problem.

Make the work easier If you need something from a team that doesn't have time to provide it, try asking them for something smaller. That might mean a single feature you absolutely need, instead of the several you'd really like. If they're blocked by dependencies of their own, think about whether there's anything you can do to help unblock them: if you can solve their problem so they can solve yours, then everyone wins! Sometimes you can end up taking a side quest down a chain of dependencies until you find a small nudge that will let everyone start moving again.

Alternatively, you might be able to offer to do part of the work for them, for example by having another team write code and send it to them for review. Be aware that this offer might not be as helpful as you intend: supporting an untrained person through making a change in a difficult codebase, for example, can often take more effort than doing it yourself. Don't be offended if the team doesn't take you up on the offer.

Get organizational support If the team you're waiting for is doing something they think is higher priority, find out whether they're right. If priorities are unclear, ask your organization's leadership to adjudicate on which project should "win." Be respectful and friendly, but ask enough questions to understand. I hope it goes without saying, but if your organization considers your work to be *less* important, you should leave the other team alone to focus on the thing they're doing.

But if your blocked project is genuinely more important, this may be a case for escalation to a mutual leader. No matter how frustrated you are, deliver the unemotional facts: explain what you need, why it's important, and what's not happening. Consider discussing the situation with your project sponsor or manager before escalating. They may have alternatives to suggest, or may be willing to have some of those conversations for you.

Warning

Escalating doesn't mean raising a ruckus or complaining about the other team. It means holding a polite conversation with someone who has the power to help, and trying to solve a problem together. Keep it constructive.

Make alternative plans If the team's really not going to be available, you'll need to find another way around the blockage. That might mean rescoping your project, choosing a different direction, or shipping later than you'd intended to. It is what it is. Make sure you talk with your stakeholders and project sponsor about any change of dates and make sure they understand what you're blocked on. They may have ideas for unblocking that haven't occurred to you.

BLOCKED BY A DECISION

Should the team take path A or path B? Should they design for a single, specific use case or try to solve a broader problem? How should they lay out their architecture, APIs, or data structures? So much depends on the details, and it's difficult to make progress without them. You could design for maximum flexibility, but that's expensive, and you want to avoid overengineering. But it will feel terrible if you have to throw your solution away in a year because you took the wrong path. So you ask your stakeholders for specific requirements, use cases, or other decisions. And you get...nothing. How can someone ask you to build something and not know what they want?!

What's going on?

When you're waiting for a decision from someone else, it's tempting to think of their work as easier than yours. They just need to decide what they want, right? Then you have the actually difficult work of building it. I've seen this bias a lot for decisions that need to come from outside engineering. "Why won't the product team just...?" As with many uses of the words "just" or "simply," the answer will be complicated. Product or marketing (or anyone else) can't make a snap decision any more than you can. Or they may be waiting for information from someone *they* depend on before they can decide.

They might also not understand what you're asking. This is especially common when you have engineers explaining an engineering problem to nonengineers. I've seen an engineer ask a stakeholder, "Do you want X or do you want

Y?" and hear "Yes, that would be great!" in response. That stakeholder isn't intentionally being obtuse! Different contexts and different domain languages mean that they don't see much distinction between the two things you've asked for, so it's impossible for them to make an informed choice.

Navigating the unmade decision

When your decision is blocked on someone else's decision, have empathy: the decision is likely not any easier where they're standing. But, of course, you don't want to wait forever, so here are some techniques for making progress when someone's having trouble deciding.

Understand and explain Remember that you're on the same side. Rather than seeing the person you're waiting for as an obstacle, see if you can navigate the ambiguity together. Understand what information or approval they need to make a decision and try to help them get it. Make sure they know why it matters and what can't happen until and unless they decide. Explain the impact on the things *they* (not you!) care about. If they're blocked, understand who or what they're waiting for.

Make the work easier Think about how you're asking the question, and build a mental model of how the other person is receiving it. Really try to get into their head: if you were them, how might you interpret the words? Are there easy misunderstandings? Think about whether you can use pictures, user stories, or examples to reframe the question. The decision might be easier once everything clicks for them.

Sometimes the problem is that the decision doesn't have a clear owner and the various stakeholders just can't agree. If the decision you need is blocked by conflict, consider playing mediator, helping each side understand the other's point of view and find a solution that suits them both. If your decision makers are blocked by *their* decision makers, give them the information they need in the format they need to take back. Take the time to give them talking points that the people *they'll* need to interact with will understand. And if some parts of the decision are more important than others, explain what's hard or expensive to reverse later and what doesn't really matter.

Get organizational support If you're still not able to get a key decision made, talk to your project sponsor about what they'd like to do. They may have some ideas about paths forward, or be in some rooms you're not in where they can push for a decision to happen.

Make alternative plans Moving a project ahead with a major decision still unmade tends to be unsatisfying and complicated: when you have to stay open to all kinds of future directions, the solutions can't be as sleek and elegant. They're often more expensive too. But sometimes keeping your options open is the best choice you have available.

Alternatively, sometimes it's OK to make your best guess about the right path and take the risk that you're wrong. If you do guess, document the trade-offs and the decision.[1] Make sure the decider knows that you're guessing and understands the implications of the direction you chose. Spend some time thinking about the worst that could happen and what you might later wish you'd done, and mitigate those risks in any way you can.

Finally, be realistic. If your organization can't make this crucial decision, is working around it going to be enough? Will you just run into the same problem for the next decision? It may be time to accept that you just can't continue the work for now. If that's the case, talk to your project sponsor, tell them what you've tried, and make sure they agree that it's not possible to proceed.

BLOCKED BY A SINGLE $%@$% BUTTON CLICK

We've all been there, and it's incredibly frustrating. You're waiting on a team or an approver that just needs to check a box, deploy a config, or review a five-line pull request. It'll take them 10 minutes! Why don't they just click the freakin' button?!

What's going on?

I used to be on a team that configured load balancing for everyone else at the company. Our documents said we needed a week's notice to add a new backend to our balancers but, truthfully, each request took about half an hour. We needed to provision extra capacity, change configurations, and restart services. It was a frequent task, and a straightforward one for someone who knew what they were doing. Other teams knew that what we were doing wasn't rocket science, so we regularly got requests like "We're launching tomorrow, so please set up our load balancing today." And, usually, we didn't.

Why did we insist on a week's notice? Because these configurations weren't our only work. Hundreds of teams used our banks of balancers, and load

---

1 Consider Lightweight Architectural Decision Records (*https://oreil.ly/BO1Kq*) for showing why you made the choice you did.

balancing was just one of the four critical services my team supported. We didn't want to react constantly: we wanted to plan out our weeks, and to make these configuration changes in batches rather than continually restarting services every time. As a result, we had little sympathy for people who came in hot and angry about why we hadn't done the thing they'd told us about only a few hours ago. Our team motto became "lack of planning on your part is not an emergency on mine."[2]

Once again, the world looks very different when you're on the other team! Your request is only one small part of someone's work, one little block on their time graph. It may be only a single button click for them, but they have a lot of other people sending them unclicked buttons too. They might be understaffed and struggling. They might be finding edge cases as they work to improve their operations. I've seen teams realize that their process for handling requests can't scale, and dedicate some team members to building a better approach, making the team short-staffed and even slower in the short term.

Bear in mind that other teams often take on accountability when they give approval. When you ask a security team to "just click the button" to approve your launch, or a comms team to approve your external messaging, you're asking them to share (or entirely own!) responsibility if something goes wrong. They shouldn't do that lightly.

Navigating the unclicked button

If the team is just not doing the work, then you can use most of the mechanisms for handling blocked dependencies that I discussed earlier in the chapter. But if it turns out that there's a standard way of interacting with them and you didn't use it, you'll need a different approach. If you're not in a real hurry, maybe you can just wait and let the team get to the work when they get to it. But if you have a real deadline, you can't exactly go back in time and use the process. Here are those techniques again:

---

2 Looking back, I have much more sympathy for the other teams. The number of configurations needed to run a service in production was massive, and any team coming up to a launch had a lot to think about. Yes, they probably should have realized they needed load balancing ahead of time, but we were one of 15 prelaunch conversations they needed to have. And we should have figured out a way to replace our manual steps with something self-service for the most common cases. Perspective.

Understand and explain If you really need to skip the queue, try asking. Be as polite and friendly as possible: you'll get better results by apologizing rather than yelling at the busy team.

If someone goes out of their way to help you, say thank you. In companies that have peer bonuses or spot bonuses, there's already a structure for saying thank you: use it. If the team is located together, can you send them a thank-you gift, like fancy tea or chocolate? At least include them in the list of people you thank in the launch email. Afterward, you won't be remembered as the team that asked for something at the last minute, you'll be remembered as the team that built bridges and made friends.

Make the work easier Just like with a team dependency, make the thing you need as small as possible. Structure your request so it's easy to say yes to, with as little reading needed as possible. (See Figure 6-3 for an example.) If you have access to other kinds of requests the team gets, look at what problems tend to arise: what information is missing, what's complicated, what's controversial. Try to make your ticket be one of the easy ones, the kind that the person doing the work picks up first because they won't need to think too hard about it.

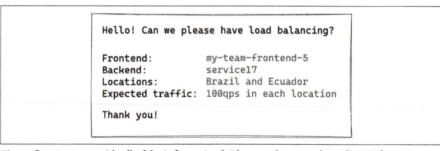

Figure 6-3. A request with all of the information laid out and not much reading to do.

Get organizational support If you *really* can't get help any other way and the deadline matters, you may need to ask for help skipping the queue. Be warned: escalating may build bad blood with the team you're looking for help from: nobody likes having their director or VP asking them to move a request to the front of the line. Mitigate any potential animosity as much as possible by being clear that you understand why the team has this process, and that you're apologetically asking to circumvent it just this once.

Make alternative plans It's possible that having (or making) a connection with someone on that team will be enough to raise the priority. (This isn't how the world *should* work, but often it's how it *does* work.) If none of these options do work, though, you may just be waiting. Let your stakeholders know you'll be delayed.

BLOCKED BY A SINGLE PERSON

If waiting for a team is frustrating, it can be even worse when a whole important project is waiting on a *single person*. The work is allocated, it's on someone's desk, and they're just not doing it.

What's going on?

I once had a project blocked by a coworker who needed to write a couple of short Python scripts to solve a problem. Weeks passed, and the scripts didn't materialize, and my urge to just write them myself got stronger every day. My colleague always had a good excuse: there was an outage, or he'd needed the day off, or his computer had a hardware problem that needed to be fixed. After a few weeks of being frustrated at his lack of progress, I realized he wasn't slacking: he was intimidated. He'd come from an operations role and had been used to the kind of interrupt-driven work where you bounce from fire to fire, rarely getting a block of focus time. This project was his opportunity to begin writing code, but he didn't really believe he could do it, and so he couldn't get started.

The reasons your colleague gives you for being blocked aren't necessarily the real ones. The person could be intimidated, stuck, or oversubscribed. They could be stressed out about something in their personal life that makes it impossible to focus, or they might be getting messages from their leadership about what's most important—and are nervous to tell you that your project isn't on that list. Or maybe they didn't understand what you were asking for the first two times you explained it, and they're too embarrassed to ask a third time. All of these causes look the same from the outside: the person's just not doing the work.

Navigating a colleague who isn't doing the work

When someone is having trouble getting their work done, there are some ways you can help. Let's be clear: you're not this person's boss or therapist, and it's not your role to fix their procrastination. But you may be able to get the outcome you want and, if it's a less experienced person, take the opportunity to teach them some skills. Here's what you can do:

Understand and explain See if you can learn more about what's going on with your colleague. Next time you talk with them, don't just accept their promise that the work will be ready in another week: dig a little deeper. You might not be able to find out what's really going on, but you'll get a sense for whether it's something you'll be able to help with.

Be very clear about why the work is necessary. This is particularly useful when you're waiting on a more senior colleague, who (presumably!) is making deliberate judgments about relative importance and not just procrastinating. Describe the business need, and show what can't happen until their work happens. If they have a heavy workload, ask them to let you know if they won't be able to get the work done, so you'll have time to find an alternative. Consider setting an earlier go/no-go deadline after which you'll both understand that they're not going to be ready in time and you should find an alternative approach.

Make the work easier Make it as easy as possible for the person to do the thing you want. That might mean adding structure, breaking the problem down, or creating milestones: when the project is too difficult, sometimes even thinking about how to break it up can be too difficult.[3] Don't obfuscate the request by making the person search through a document for action items assigned to them or intuit what you're asking them to do: just spell out what you need. Look at Brian Fitzpatrick and Ben Collins-Sussman's "Three Bullets and a Call to Action" technique for asking for something from a busy executive: it works here too!

What Are You Asking For?

I love the "Three Bullets and a Call to Action" method that Brian Fitzpatrick and Ben Collins-Sussman outline in their book, *Debugging Teams* (O'Reilly). As they write: "A good Three Bullets and a Call to Action email contains (at most) three bullet points detailing the issue at hand, and one—and only one—call to action. That's it, nothing more—you need to write an email that can be easily forwarded along. If you ramble or put four completely different things in the email, you can be certain that they'll pick only one thing to respond to, and it will be the item that you

---

3 If you're the procrastinator, consider the calendaring trick I mentioned in Chapter 4: put the task you need to do in your calendar. If even understanding the first step is difficult, make a calendar block just for figuring out what the first step of the work will be, and schedule a separate block for working on that step. Give Future You the smallest possible tasks to do.

care least about. Or worse, the mental overhead is high enough that your mail will get dropped entirely."

If your colleague is overwhelmed, this may be an opportunity for coaching. Reassure them that what they're working on is legitimately difficult but learnable. Help them, but try not to take over.[4] Ask questions, answer questions, and help them find their way.

If your colleague seems willing to do the work but is having trouble getting started, see if you can work with them on it. This was the solution for the colleague I mentioned earlier: pairing on the scripts got him past the intimidating first steps of the work and able to continue on his own. Pairing can also take the form of whiteboarding together or sitting down together to edit a document at the same time. This last one can sometimes be a good approach when you're waiting on your manager: you can meet for a one-on-one meeting, suggest that they use the time to do the thing you need, and stay there with them so you're available to answer any questions they have along the way.

Get organizational support While you should try other approaches before escalating, ultimately someone is not doing their job, and that's a people management issue. Having a difficult conversation with their manager is uncomfortable, but if the other person is the reason a project is going to fail, you're not doing your job either if you ignore it. Just like I've said for other situations, escalating doesn't mean complaining: it means asking for help. If your colleague is blocked because they're working on something their manager thinks is more important, for example, talking to that manager may be the only way to adjust the priorities.

BLOCKED BY UNASSIGNED WORK

What about when the work isn't assigned to anyone? When a group of teams are working together to solve a problem, sometimes there's an effort that everyone agrees needs to happen but that isn't on anyone's roadmap. It's too big for any one engineer to tackle, it needs a lot of dedicated time, or it will involve creating a new component that will need ongoing ownership and support: it needs a team

---

4 It is *really* difficult to watch work that you care about being done badly, but try not to step in and do it for them. If it's critical that the problem gets solved right now, see if you can work with the other person and get them to take each step, rather than just doing it yourself. Your colleague won't learn to drive if you take the wheel.

to own it. Maybe there are several different teams that consider themselves *involved* in the work—they'd turn up to a meeting about it and have many opinions on any RFC—but none of them intend to commit code to achieve it.

What's going on?

This is an example of a *plate tectonics* problem (see Chapter 2). Every team has clear boundaries on what they're responsible for. Strong boundaries can be great —they keep everyone focused—but now there's some crucial foundational work that doesn't belong to anyone. We saw this situation in the SockMatcher scenario in Chapter 3. Many engineers cared about the architecture: Geneva's working group to discuss the problems drew many participants. But the challenges were too big for anyone to solve as a side project. Unless it gets dedicated attention, the work is just not going to happen.

If nobody is assigned to do the work, there's limited value in breaking it down, optimizing it, or making plans: you're stuck until someone owns it. It's very common for engineers to keep trying to solve an organizational blocker like this by putting more effort into the adjacent technical work. (See Figure 6-4 for an illustrative treasure map: that impassable ridge is your lack of staffing!)

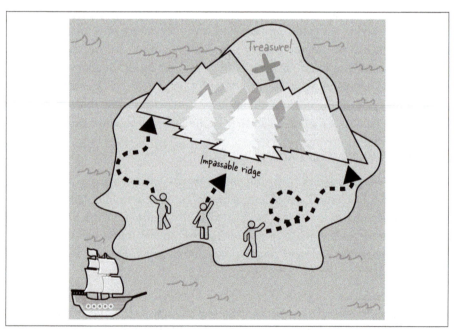

Figure 6-4. Unless you can find a way past the impassable ridge, it doesn't really matter which of the first three paths you take. But teams invest a lot of time debating which doomed path to take.

But unless the organizational problem is solved, no amount of designs and clever solutions will help. The organizational gap is the crux of the problem. If you can't solve that, you're wasting your time.

Navigating the unassigned work

When you're blocked on work that doesn't seem to be anyone's responsibility, there are a few things you can do.

Understand and explain It's not always obvious that work is unowned, particularly when there are a lot of teams that are closely adjacent to it. Since you're the person most interested in its success, don't be surprised if other people think that *you're* the owner now. Have a whole lot of conversations, follow the trail of clues, and figure out exactly what's going on. Interpret the information and draw explicit conclusions. Then write them down! (See the following sidebar.) Keep it brief and make clear statements: just like I said about design documentation in Chapter 5, wrong is better than vague, and you won't uncover any misunderstandings unless you remove the ambiguity.

Rollup

Denise Yu, a manager at GitHub, describes "the art of the rollup" (*https://oreil.ly/5U1wf*): summarizing all of the information in one place to "create clarity and reduce chaos." It's a versatile technique, useful in any situation where there's a ton of backstory and several different narrative threads and where some people might not have kept track of what's going on.

Aggregating the facts that go into the rollup is a great way to build your knowledge and make sure you understand what's going on. But writing it all down also might mean you synthesize new information that nobody had articulated before. Perhaps Alex says, "The new library will give us authentication for this end point." And later Meena tells you, "We won't be able to upgrade to the new version of the library until Q3." You can write down both facts, but also the interpretation, "We won't have authentication until at least Q3."

> That conclusion might seem obvious with all of the context, but if nobody has reached it before, it may come as a shock to some. By spelling it out, you give everyone an opportunity to react to the information and course-correct.

Make the work easier If you have time to mentor, advise, or join the team doing the work, mention that in your rollup. Directors may be more inclined to build a team around an existing volunteer, and the chance to learn from a staff+ engineer can be an incentive that they can offer to other team members they're inviting to join.

Get organizational support Your highest-value work on this project will be advocating for organizational support and an owning team. Before you set out to find a sponsor for the work, make sure you've honed your elevator pitch and can explain why the problem is worth your organization's time. You'll want your sponsor not just to believe you, but to be able to justify the effort to their peers and leadership if they need to.

Make alternative plans If there's been a credible promise of staffing, be patient and give it time to happen: project staffing involves readjusting teams, which managers and directors (understandably) prefer to be deliberate about. But if you can't get a commitment for anyone to own the work, or if it's always "next quarter" without any specifics about where the staffing will come from, that's a sign that your project isn't a high priority for the organization and should be postponed.

BLOCKED BY A HUGE CROWD OF PEOPLE

The last type of roadblock I'm going to talk about is when the project needs help from everyone: it's the kind of deprecation or migration where *all* of the teams using a service or component need to change how they work. Chances are, not all of them will want to.

I've worked on many, many software migrations, and I know how frustrating they can be. You have good reasons for moving off an old system, you know exactly what needs to happen and you've communicated plenty, but other teams are ignoring your emails. When you chase them, they say they're busy—but you're busy too! Why can't they see that this work matters?

What's going on?

Every team and every person has their own story. One might be broadly in favor of the migration but just doesn't have time. Another might be opposed to the change: they might prefer the old system, or feel that the new one is missing some feature they care about. A third group might not have feelings either way, but they're just tired of a constant, demoralizing stream of upgrades, replacements, and process changes that they don't seem to get any benefit from.

The people pushing *for* the migration and the people pushing *against* are all being reasonable. But being stuck in a half-migration situation isn't good for anyone: teams need to spend time on supporting both the old system and the new system, and new users may have to spend time understanding which to use, particularly if the migration stalls and seems at risk of being canceled.

Navigating the half-finished migration

The half-migration slows down everyone who has to engage with it. This is a place where a staff engineer can step in and have a lot of impact. Here are some ways you can get everyone over the finish line.

Understand and explain The narrative of a migration can get lost, and I've sometimes heard a migration framed as "that infrastructure team just wants to play with new technology" when the reality is an immovable business or compliance need that infrastructure team doesn't love either. Equally, I've heard "that product team doesn't care about paying down technical debt; they only want to create new features," when the team in question has already spent half of the quarter reacting to other migrations and are desperate to get started on their own objectives. Understand both sides and help tell both stories. Show why the work is important and also why it's hard. Be a bridge.

Make the work easier Once again, the key to convincing other people to do something is to make the thing easy to do. In general, lean toward having the team that's pushing for the migration do as much extra work as possible, rather than relying on every other team to do more. If there's any step you can automate, automate it. If you can be more specific about exactly what you want each team to do—which files you want them to edit, for example—spend extra time providing that information. Also, it hopefully goes without saying, but the new way has to *actually work* without forcing the teams to jump through hoops.

Try to make the new way the default. Update any documentation, code, or processes that point people toward the old path. Identify people who might

encourage the old way and ask for their help. If you have the organizational support for it, consider an allow-list of existing users, or some friction to make the
old way harder to adopt. One approach I've seen is to keep the old way working
fine so long as new users write "i_know_this_is_unsupported" or similar as part
of their configuration.

Show the progress of the work. I worked on one project where I needed to
get hundreds of teams to change their configs to use a different endpoint. When
I shared the graph showing how many had been done and how many were left to
do, people were more eager to do their part to get the numbers down. Something
in our brains really likes seeing a graph finish its journey to zero![5]

Some teams will be genuinely too busy to move, or will have a use case that
isn't quite supported by your automation. There might also be some components
that nobody owns anymore, so there's no team to update them. Can you pair with
the team, or make the change for them?

Get organizational support The migration will be easier if you can show that
this work is a priority for your organization. The corollary to this, of course, is
that the work *should* actually matter. If teams are snowed under by changes, your
organization should be prioritizing the most important ones and finishing the
first set of changes before starting the next set. If you can't convince your leadership that your migration should be reflected on the organization's quarterly goals
or list of important projects, maybe it's a sign that this is not where you should
be spending time right now.

Make alternative plans By the end of the migration, you may need some creative
solutions. If you have an organizational mandate and teams are still not moving,
can you withdraw support for the old way, or even start adding friction along the
path, for example by introducing artificial slowness or even turning it off at intervals? Don't do this without strong organizational support (and don't do it at all if
it's going to break things for your customers; be sensible). If the final teams are
really refusing to move off the old system, can you make them its sole supporters
and owners, so that it runs as a component of their own service? Last out turns
out the lights, friends.

---

5 This only works if the graph is showing progress: if it's clear no one is doing the work, it can backfire. But
the social side of influencing people is such a great tool: "Everyone else is doing it, why aren't we?"

You're Lost

Let's move on to the second set of reasons you might have trouble making progress: you're just lost. It's not that you're blocked by anything that you can see: you just don't know the way. This might be because you don't know where to go now, because the problem is too hard, or because you're not sure if you still have organizational support for what you're doing. You'll use different techniques in each case.

YOU DON'T KNOW WHERE YOU'RE ALL GOING

Imagine 40 teams working on the same legacy architecture. There's a single massive codebase, a decade of regrettable decisions, a tangled mess of data that's owned by everyone all at once. Teams are scared to refactor existing code, so they bolt new features onto the outside. You've been tagged as the leader who is responsible for fixing it and you have a team dedicated to the work. There's an organizational goal to "modernize the architecture." It feels like you have a clear mandate to do the work! And yet...there are so many decisions to make. There are so many stakeholders, a huge number of comments on every document, too many voices in every meeting. It's been a few months and you've made no real progress.

What's going on?

We saw this with the SockMatcher scenario in Chapter 3: it's difficult to solve a problem that half the company cares passionately about! Everyone has an opinion. Everyone's sure they know the right thing to do. In this example, there's almost certainly a group of people advocating for creating microservices. Another faction might want feature flagging and fast rollbacks so that it's safer to make changes. A third wants to move to event-driven architecture. A fourth doesn't care what happens so long as the underlying data integrity problems are resolved. Those are just four of the many fine suggestions.

A huge group tackling an undefined mess will almost inevitably get stuck in analysis paralysis. Everyone agrees that *something* should be done, but they can't align on what. You can't steer this project because it's just a bunch of ideas: there's no project to steer.

Choosing a destination

You can't start finding a path to the destination until you're very clear about what that destination is. Here are some approaches for choosing it:

Clarify roles With a group this big, the leader can't be just one voice in the room. Be clear about roles from the beginning. Be explicit that you want to hear from everyone, but that you're not aiming for complete agreement: you will ultimately make a decision about which direction to take. If you don't feel that you have the power to name yourself the decider, ask your project sponsor or organizational lead to be clear that they will back you up. If you don't have that kind of organizational support, you may not be set up to succeed.

Choose a strategy Unless you know where you're going, you have little chance of getting there. Set a rule that, until you all agree on exactly what problem you're solving, nobody is allowed to discuss implementation details.[6] If you can, choose a small group to dig into the problems and create a technical strategy. Emphasize that any strategy will, by definition, make trade-offs, and that it can't make everyone happy. You'll pick a small number of challenges, and leave the other real problems unsolved for now. Chapter 3 has lots more on how to write a strategy: set the expectation that it will not be a short or painless journey.

Choose a problem If you've been assigned engineers who are eager to start coding in the service of the goal, it can be frustrating (and politically unpopular) to say that you need to spend time on a strategy first. If you really don't have time to evaluate all of the available challenges and rank the work by importance, choose *something*, any real problem. Set expectations that you won't allow the group to get diverted by the other (very real) issues, but that you fully intend to return to them after solving the first one. Once again, wrong is better than vague: any deliberate direction will probably be better than staying frozen in indecision.

Choose a stakeholder One way you can choose a problem to solve is by choosing a stakeholder to make happy. Rather than solving "the shared datastore stinks and we need to rethink our entire architecture!" can you solve "one team wants to move its data elsewhere"? Reorient the project around getting *something* to *someone*. Aim to solve in "vertical slices": first you help one stakeholder complete something, and then another. Progressing in *some* direction can help break the deadlock and clarify the next steps. Once you're showing some results, consider revisiting the idea of creating a strategy and having a big-picture goal.

---

6 This will never entirely work, but keep trying. The more you can keep people out of the weeds, the more chance you have of succeeding.

YOU DON'T KNOW HOW TO GET THERE

What if you know exactly where you're going? The destination is well understood and there aren't blockers in your path, but you're still not getting there. You're not sure how to solve the next problem in front of you, or the project is huge and you're not even sure which problem you're supposed to solve next. I mentioned last chapter how self-fulfilling imposter syndrome can be: if you're finding the work difficult, that can become a vicious cycle that makes you have less capability for tackling it. Maybe you're avoiding thinking about it, but the longer you ignore the project, the worse it feels.

What's going on?

The project is just difficult! It might be that there are a massive number of topics to keep track of and you feel entirely out of your depth, particularly when something is going wrong. Or, there's one impossible task, a technical challenge or an organizational hurdle, and you just don't know how to get past it. If you haven't seen this kind of problem before, it might take you time to even recognize what's happening, and longer to find solutions.

Finding the way

The path forward is *unknown* but not *unknowable*! Here are some techniques to start finding your way:

Articulate the problem Make sure you have a crisp statement of what you're trying to do. If you're struggling to articulate it, try writing it out or explaining it out loud to yourself. Notice places where you could be more precise about who or what you're referring to or what *is* happening versus what *should be* happening. Refine your understanding by talking through the problem with anyone who's willing to discuss it with you.

Revisit your assumptions Is it possible that you've already assumed a specific solution and are struggling to solve the problem only within that context?[7] Are you looking for a solution that's an improvement on every axis when trade-offs might be acceptable? Are you dismissing any solutions because they seem "too

---

7 As Leslie Lamport cautioned (*https://oreil.ly/LO1jT*), you should "specify the precise problem to be solved independently of the method used in the solution." He wrote, "This can be a surprisingly difficult and enlightening task. It has on several occasions led me to discover that a 'correct' algorithm did not really accomplish what I wanted it to."

easy"? Explaining out loud why you think the problem can't be solved might help you discover some movable constraints you hadn't noticed before.

Give it time Have you ever been blocked by a coding or configuration problem that you just couldn't crack, and then the next morning you could immediately solve it? Sleep is amazing. Vacations can do the same thing. I've found that if I take a few days away from a problem, I'll almost always come back with better ideas, even if I haven't thought about it in the meantime.

Increase your capacity Trying to solve a problem in the tiny spaces between meetings will constrain the ideas you can have. Schedule yourself some dedicated time to really unpack the situation in your mind: it can take a few hours even to clear out the noise of whatever else you were thinking about. Aim to bring your best brain to that meeting with yourself: for me that means good sleep, non-stodgy food, plenty of water, and a room with good light and air. You know your own brain: do whatever makes you smart.

Look for prior art Are you really the first person to ever solve a problem like this? Look for what other people have done, internally and externally. Don't forget that you can learn from domains other than software: industries like aviation, civil engineering, or medicine often have well-thought-out solutions to problems tech people think we're discovering for the first time!

Learn from other people Talking through the problem with a project sponsor or stakeholder can sometimes give you enough extra context or ideas to find your next steps. You can also learn from people outside your company. Most technical domains have active internet communities. See if there's a place where experts on the topic hang out, and spend time there absorbing how they think and what keywords and solutions they mention that you don't know about.

Try a different angle Spark creative solutions by looking from another angle. If you're trying to solve a technical problem, think about organizational solutions. If it's an organizational problem, imagine how you'd approach solving it with code. What if you had to outsource it: who would you pay to solve this problem and what would they do? (Might that be an option?) If you weren't available, who would this work get reassigned to and how would they approach it?

Start smaller If you're overwhelmed with tasks and it's not clear what comes first, try solving one single small part and see if you can feel a sense of progress and make the rest of the work feel more achievable. Another angle is to ask yourself whether you really need to solve the problem *well*. Could a hacky solution be good enough for now? Or can you start with a terrible solution and iterate so that you're not starting with a blank page?

Ask for help While it might feel like your skills are the only thing standing between success and failure, you're not alone. Ask for help from coworkers, mentors, or local experts. If you're someone who hates asking for help, remember that by learning from other people's experiences, you're amortizing the time they had to spend learning the same thing: it's inefficient to have you both figure out solutions from first principles.

YOU DON'T KNOW WHERE YOU STAND

Here's the last form of being lost and, in some ways, it's the scariest: you don't know if your work is still necessary. A comment you heard at the latest all-hands meeting might make you nervous that a new initiative will derail yours. Maybe you've noticed that your manager or project sponsor is checking in less often, and seems less interested in your results. Or there was a company announcement that listed all of the important projects—and yours wasn't on it. Yikes. Some of the people you're working with seem disengaged: they talk about your project more as "if" than "when" and they're prioritizing other work. Have they heard something you haven't? Is the project still happening? Nobody is telling you anything. Are you still in charge?

What's going on?

Organizational changes, leadership changes, and company priority changes can all affect enthusiasm for your project. If you have a new VP or director, they may not think the problem you're solving is important, or, sometimes worse, they may think it's so important that they're solving it with a much bigger scope than you had taken on and a different lead. Your project genuinely could be at risk of being killed, and nobody's thought to tell you. This missing communication is especially common if your leadership position is an unconventional one—for example, if you're in an organization adjacent to the one that's doing most of the work, or if you're leading people who are more senior than you are. It's easy to be forgotten when priorities change, or left out of meetings where decisions are happening.

Or it might not be that at all! Your project might be going so well that you're just getting less attention from your leadership. When there are fires elsewhere, nobody checks in on a project that's humming along. Silence could mean "keep doing what you're doing."

Getting back on solid ground

Continuing on the same path without knowing where you stand is just a recipe for stress, and you may be wasting your time. Here are some things you can do:

Clarify organizational support Brace for the idea that you might not like the answers, and go find out what's happening. Talk with your manager or project sponsor, explain what you've heard, and ask whether your project is intended to continue.

Clarify roles If you're the lead but find yourself hesitant to claim that title, or if you aren't sure what you're allowed to do, you need to formalize roles. The RACI matrix I described in Chapter 5 might be a useful tool here, as might the role description document from Chapter 1. By the way, if you're trying to run a project with a title like "unofficial lead," that's an invitation to fail—if you're the lead and nobody else knows you are, you're not the lead.

Ask for what you need If you're missing the authority, the official recognition, or the influence to do the work, who can help you find those things? It's natural to want some reassurance that your project is still important. It's fine to ask for a mention of the project at an all-hands meeting or to have it listed on the organization's goals. You might not get what you want, but you definitely won't if you don't ask.

Refuel It's demoralizing to work on something that nobody else seems to care about. If you and your team are feeling low on energy, you may need to deliberately build that back up again. Refueling could mean setting new deadlines, building a new program charter, or having a new kickoff or off-site meeting: resetting with "Welcome to Phase 2 of the project" is somehow more motivational than "Let's keep doing what we've been doing, but I swear it will be different this time." If you can add a new team member or two who are raring to go, their enthusiasm can be enough to get the team moving again.

You Have Arrived...Somewhere?

There's a third reason projects stop: the team thinks the project has reached its destination...yet somehow the problem is not solved.

I've seen many projects end like Figure 6-5: just short of their goal. All of the tasks on the project plan are completed, the team members have collected their kudos and moved on to other things, yet the customer is still not happy.

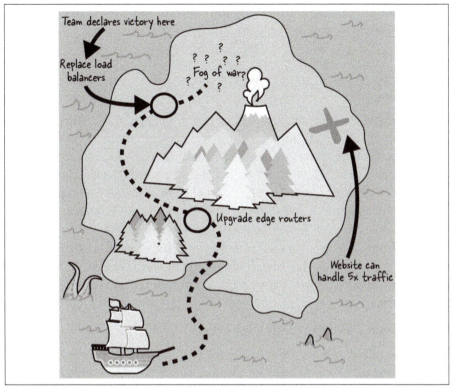

Figure 6-5. The team declared victory and went home—but there was another, better treasure they never got to.

In this section we'll look at three ways you might declare victory without actually reaching your destination: doing only the clearly defined work, creating solutions you don't tell your users about, and shipping something quick and shoddy. Watch out for these end states!

BUT IT'S CODE COMPLETE!

I feel like I've had this conversation a hundred times in my career:

> *"I'm excited for the new foo functionality. When will it be available?"*
>
> *"Oh, it's done!"*
>
> *"Amazing! How do I start using it?"*
>
> *"Well..."*

The "Well..." is always followed by the reasons I can't use it *today*. Foo still has hardcoded credentials; it's only running in staging, not in production; one of the PRs is still out for review. But it's *done*, the other person will insist. It's just not *usable* yet.

What's going on?

Software engineers often think of their job as writing software. When we plan projects, we often only list the parts of the work that are about writing the code. But there's so much more that needs to happen to allow the user to *get* to the code: deploying and monitoring it, getting launch approvals, updating the documentation, making it run for real. The software being ready isn't enough.

Heidi Waterhouse, a developer advocate for LaunchDarkly, once blew my mind with the observation that "nobody wants to use software. They want to catch a Pokémon." A user who wants to play a video game doesn't care what language the code is in or which interesting algorithmic challenges you've solved. Either they can catch a Pokémon or they can't. If they can't, the software may as well not exist.

Making sure the user can catch a Pokémon

As the project lead, you can prevent work from falling through the cracks by looking at the big picture of the project, not just which tasks were done. Here are some techniques you can use:

Define "done" Before you start the work, agree on what the end state will look like. The Agile Alliance proposes (*https://oreil.ly/3Mray*) setting a *definition of done*, the criteria that must be true before any user story or feature can be declared finished.[8] This might include a general checklist for all changes—I've often seen PR templates include a section for explaining how the change was tested, for example—as well as a specific set of criteria for individual projects. Similarly, user acceptance testing lets intended users of a new feature try out the tasks they'll want to do with it and agree that those features are working well. Even internal software can have user acceptance tests. Until those are completed and the user declares that they're happy, nobody gets to claim the project is done.

---

8 They also distinguish between "done" and "done-done," and credit a 2002 article from author and agile coach Bill Wake (*https://oreil.ly/2Tsnm*) where he asks the enigmatic question, "Does 'done' mean 'done'?"

Be your own user Is it possible for you to regularly use what you're building? Of course, this isn't always going to apply, but if there's a way for you to share your customers' experience, take the time to do that. This is sometimes called eating your own dog food (*https://oreil.ly/qHyWM*) or "dogfooding."

Celebrate landings, not launches Celebrate shipping things to users, rather than milestones that are only visible to internal teams. You don't get to celebrate until users are happily using your system. If you're doing a migration, celebrate that the old thing got turned off, not that the new thing got launched.

IT'S DONE BUT NOBODY IS USING IT

Have you ever seen a platform or infrastructure team spend months creating a beautiful solution to a common problem, launch it, celebrate, and then get frustrated that nobody seems to want to use it? They're sure it's better: it genuinely improves its users' lives. But teams are still doing things the old, difficult way.

What's going on?

The team is not thinking beyond the technical work. Unfortunately, internal solutions are often marketed like this: we create some useful thing. We give it a cute name. We (maybe) write a document explaining how to use it. And then we stop. The thing we created probably has potential users who would love it, but they have no way to know it exists; if they stumble across it, the name offers no hint as to what it does. Common search terms for the problem don't suggest this solution. I call these "Beware of the Leopard" projects, from one of my favorite parts of Douglas Adams's classic book *The Hitchhiker's Guide to the Galaxy* (Pan Books).

> *"But the plans were on display..."*
>
> *"On display? I eventually had to go down to the cellar to find them."*
>
> *"That's the display department."*
>
> *"With a flashlight."*
>
> *"Ah, well, the lights had probably gone."*
>
> *"So had the stairs."*
>
> *"But look, you found the notice, didn't you?"*

"Yes," said Arthur, "yes I did. It was on display in the bottom of a locked filing cabinet stuck in a disused lavatory with a sign on the door saying 'Beware of the leopard.'"

It's like the creators of the solution are trying to hide it! The information *exists*—someone got to mark their documentation milestone as green—but it will never be found by anyone who doesn't know what they're looking for.

Selling it

Michael R. Bernstein has a great analogy (*https://oreil.ly/u65tj*) for creating solutions and then not marketing them at all. He says it's like a farmer planting seeds, watering, weeding, and growing a crop, and then just leaving it in the field. You need to harvest what you grew, take it to people, and show them why they want it. The best software in the world doesn't matter if users don't know it exists or aren't convinced it's worth their time. You need to do the marketing.

Tell people You don't just need to tell people that the solution exists: you need to *keep* telling them. A lot of migrations stay at the half-migration stage because engineers assume that users will just come find the software. Help them find it. Send emails, do road shows, get a slot at the engineering all-hands. Offer white-glove service to specific customers who are likely to advocate for you afterward. If you're in a shared office, consider putting up posters! Get testimonials. Understand what anyone might be wary of, or unenthusiastic about, and make sure your marketing shows that you've thought about (and hopefully fixed) those problems. Be persistent and keep telling people until they start telling each other.

Make it discoverable Whatever you've created, make it easy to find. This means linking to it from anywhere its intended users are likely to look for it. If you have multiple documentation platforms, make sure a search on any of them will end up at the right place. If your company uses a shortlink service, set up links for all of the likely names, including the misspellings and any hyphenations people are likely to guess.[9]

---

9 Google used to have a project called Sisyphus, a name that's memorable to some but an unlikely series of letters to others. I'll always be impressed with whoever set up the shortlinks go/sysiphus and go/sisiphus as redirects to go/sisyphus. It's a good security practice too; it prevents someone standing up a fake service at the misspelled place.

IT'S BUILT ON A SHAKY FOUNDATION

The last type of "done but not *really* done" I'm going to talk about is one that can cause a lot of conflict. It's when a prototype or minimum viable product has gone into production and the users can use it pretty well, but everyone knows that it's hacked together. The user can catch a Pokémon, and the job is done: time to move on to the next thing, say the product managers! But the engineers know that the infrastructure won't scale, the interfaces aren't reusable, or the team is pushing appalling technical debt into the future.

What's going on?

There might be good reasons to ship something as cheaply and quickly as possible. When there's a competitive market—or a risk that there's no market at all—it's often more important to get *something* launched than for that something to be solid. But when the team has moved on, that cheap solution remains in place. The code may be untested—or untestable. The feature may be an architectural hack that everyone else now needs to work around.

I used to work in a data center, a long time ago, and one thing I learned there is that there's no such thing as a temporary solution. If someone ran a cable from one rack to another without neatly cabling and labeling it, it would stay there until the server was decommissioned. The same is true for every temporary hack: if you don't have it in good shape by the end of the project, it's going to take extraordinary effort to clean it up later.

Shoring up the foundations

While you can get away with shipping shoddy software in the short term, it's not a sustainable practice. You're pushing the cost of the software to yourself in future. As a staff engineer, you have more leverage than most. Here are some ways you can advocate for keeping standards high.

Set a culture of quality Your engineering culture will be led by the behavior of the most senior engineers. If your organization doesn't have a robust testing culture, start nudging it in the right direction by being the person who always asks for tests. Or be the squeaky wheel who always asks, "How will we monitor this?" or the one who invariably points out that the documentation needs to be updated with the pull request. You can scale these reminders for the project with style guides, templates, or other levers. We'll look at setting standards and causing culture change in Chapter 8.

Make the foundational work a user story Ideally, the organization agrees that shipping solutions isn't just about features—it's about preparing for the future. There's a healthy balance of feature and maintenance work, and teams build the time for high quality into the cost of their projects. Even teams that build light-weight first versions to get early user feedback or take on technical debt to get a competitive product to market faster always return to improve whatever they've shipped.

Unfortunately, few organizations work in ideal ways. You can help by making sure the user stories for the project include any cleanup work you'll need to do. You might frame this as part of user experience (nobody's excited about a product that's flaky or falls over a lot), or as laying the foundation for the next feature. If users file related bug reports or there are action items after outages, you can sometimes use those to justify the cleanup work. Focus the conversation back on the customer's needs and show that the work has a real impact on them.

Negotiate for engineer-led time If your company doesn't have a regular culture of cleaning up as you go along, see if it's possible for you to get momentum around having regular cleanup weeks. I've heard these called "fix-it weeks" and "tech debt weeks," and I've also seen a fair amount of cleanup happen during engineering exploration time, like "20% time" or "passion project week." Another option is to set up a rotation where one person on the team is always dedicated to responding to issues and making things better. Don't get hung up on the name or whether something really "counts" as technical debt. The point is that it's a dedicated time to do work that everyone in engineering knows is necessary.

THE PROJECT JUST STOPS HERE

As you navigate the obstacles and deal with the difficult situations, you'll get closer to your destination—or discover that it isn't reachable after all. Let's end the chapter by looking at four ways the project can come to a deliberate end: deciding you've done enough, killing it early, having it canceled by forces outside your control, and declaring victory.

This is a better place to stop

It's possible to get to a point where further investment is just not worth the cost and to declare that the project is "done enough." In Chapter 2 I talked about the local maximum: a team working on something low-priority because it's the most important problem in their own area, while the team next door has five more important projects that they don't have time to get to. Is your team polishing and adding features to something that's really as done as it needs to be? It might be time to declare success and do something new. Before you do, look at the failure modes in the previous section and make sure you're not abandoning unhappy customers, bailing once the technical work is done, or walking away from a half-migration. If you're all good, congratulations on reaching the new destination!

It's not the right journey to take

One of the biggest successes I ever saw celebrated could have been considered a failure. A team of senior people worked for months to create a new data storage system. Other teams were eagerly awaiting it. But as the work progressed, the team discovered that the new system wouldn't work at scale. Rather than deny reality and search for possible use cases for what they'd created, they killed the project and wrote a detailed retrospective.

Was that a failure? Not really! Other teams were disappointed not to get the system they were hoping for. But the storage team couldn't have discovered that the solution wouldn't work without exploring it. If they'd realized that the technology couldn't work for them and pushed on *anyway*, then *that* would have been a failure.

Have you heard of the *sunk cost fallacy*? It's about how people view the investments of time, money, and energy they've already made: if you've already put a lot of time and energy into something, you're more likely to stick around and see it through, even if that's a bad idea. It can be difficult to break out of this frame of mind, but without a highly tuned sense of whether something's still a good idea, you can stay on the wrong path for a long time.

Try to notice if you're in an impossible situation; if so, bail out early. Pushing on with a doomed project is just postponing the inevitable and it prevents you from doing something more useful. Good judgment includes knowing when to cut your losses and stop. Consider writing a retrospective and sharing as much as you can about what happened. Few things are as powerful for psychological safety as saying "That didn't work. It's OK. Here's what we learned."

The project has been canceled

The company's enthusiasm for what you're all doing can change. Maybe the project has been dragging on too long or is more difficult than expected, and the benefit no longer justifies the cost. Maybe a new executive has a different direction in mind, the market has changed, or your organization is overextended and looking for initiatives to cut. For whatever reason, the project isn't happening.

Someone in your management chain takes you aside and tells you the difficult news. If you're lucky, you find out before the rest of the company does.

Let's start with the feelings, because this is a tough situation. It feels *bad*. Even if your team understands and accepts the reason for the cancellation—even if you all agree!—it's jarring to suddenly drop the plans and milestones you've built. You all might feel like the work you've done has been for nothing. If you weren't part of the decision to cancel, it can feel like a personal failure. You might feel angry, disappointed, and cheated or resent that a change that affects you so much has happened in a room you weren't in. This may be a legitimate complaint: if managers make big technical decisions but leave the technical-track folks out of the room, that might be worth a conversation. But most of the time, these decisions are made at a higher level, by people who are looking at a much bigger picture and optimizing for something different than you are.

Work through your own feelings and acknowledge them. Talk it through with your manager or your sounding board. Try to understand the bigger picture and get as much perspective as you can. Then talk with your team or subleads. Tell them what's happening, leading with the why. Give them time to talk about their own reactions to the news. Respect that they might be mad at you, the project lead, as the bearer of the bad news or the person who didn't manage to "save" the project. It's important that they hear it directly from you or another leader: don't let them find out from the gossip mill, a mass email, or the company all-hands meeting.

Give yourself and your team a little time, then shut the project down as cleanly as you can. If you can stop running binaries, turn them off; if you can delete code, delete it. If you think there's a real chance that the project will be restarted later, document what you can so that some future engineer can understand what you were trying to do. (Be realistic about the chances of this resurrection happening, though.) Consider a retrospective if there's something to learn.

It's not fair, but a canceled project can derail promotions or break a streak of great performance ratings. Do what you can to showcase everyone's work. If your team members are moving on to other projects, make a point of telling their new managers about their successes and offer to talk with them or write peer reviews at performance review time. If someone's on the cusp of promotion, emphasize that to their new lead, so they don't have to build up a new track record from scratch.

Celebrate the team's work and the experience you had together. It's sad to dissolve a team that was working well together, but look for opportunities to work with those people again.

This is the destination!

Congratulations! You have done the thing you set out to do!

Before you celebrate, double-check that you have actually reached the destination. Are your measurable goals showing the results you want? Can the user catch a Pokémon? Are the foundations solid and clean? If you're really at the end, it's time to declare victory. Take the time to mark the occasion and make your success feel special. For some teams, celebrating means parties, gifts, or time off. There might be email shout-outs or recognition at all-hands meetings. Look for opportunities to give team members (and yourself!) visibility through internal and external presentations or articles on your company blog. And consider a retrospective: they're not just for looking at things that went wrong. You can learn just as much from things that went right.

If there's an aspect of your culture that you want to enforce, like people helping each other or communicating well, highlight the ways that it showed up during the project. Shout out people who went above and beyond or demonstrated any behaviors you'd like to see more of. A successful project can be a fantastic opportunity to celebrate the aspects of your culture that you most appreciate and show others what great engineering looks like.

And in the spirit of showing your organization how to do great engineering, let's move on to Part III of the book, Leveling Up.

To Recap

- As the project lead, you are responsible for understanding why your project has stopped and getting it started again.
- As a leader in your organization, you can help restart other people's projects too.
- You can unblock projects by explaining what needs to happen, reducing what other people need to do, clarifying the organizational support, escalating, or making alternative plans.

- You can bring clarity to a project that's adrift by defining your destination, agreeing on roles, and asking for help when you need it.

- Don't declare victory unless the project is truly complete. Code completeness is just one milestone.

- Whether you're ready or not, sometimes it's time for the project to end. Celebrate, retrospect, and clean up.

PART III

Leveling Up

7

You're a Role Model Now (Sorry)

"Don't think out loud," my friend Carla Geisser warned me when I became a staff engineer. "You'll find out a month later that people are talking about your half-baked idea like it's already a project." My colleague Ross Donaldson described his own role even more starkly: "Being staff doesn't absolve you of being wrong, but it does mean you need to be careful when you open your dang mouth."

This is the blessing and the curse of a staff engineer title: people will assume you know what you're talking about—so you'd better know what you're talking about! Your work will be a little less checked and your ideas considered more credible. Rather than guiding you, people will look to you for guidance.

Most of all, you'll be a role model. How you behave is how others will behave. You'll be the voice of reason, the "adult in the room." There will be times when you'll think "This is a problem and someone should say something"...and realize with a sinking feeling that that someone is *you*. The behavior you model will show your less experienced colleagues how to be a good engineer. Later, in Chapter 8, we'll look at how to *actively*, deliberately influence your organization and colleagues for the better. But this chapter is about *passive* influence, the kind that you have just by the way you act as an engineer and as a person.

What Does It Mean to Do a Good Job?

Your company might have a written definition of what good engineering means: written values, perhaps, or engineering principles. But values are what you do: the clearest indicator of what the company values is what gets people promoted. No matter how much your organization claims to encourage collaboration and teamwork, that message will be undermined if any staff engineers get to that level through "heroic" solitary efforts. If your engineering principles describe a culture of thorough code reviews, but senior engineers approve PRs without reading them, everyone else will rubber-stamp code reviews too. The work that you do is implicitly the type and standard of work that others will see as correct and emulate.

Engineering goes beyond what you do when you're talking to computer systems; it's also about how you talk to humans. So sometimes being a good engineer boils down to being a good colleague. If you're mature, constructive, and accountable, you're telling your new grads that's what a senior engineer does. If you're condescending, impossible to please, or never available, *that's* what a senior engineer does, too. You shape your company every day, just by how you behave.

BUT I DON'T WANT TO BE A ROLE MODEL!

Being a role model is not always comfortable. But as you become more senior, it's one of the biggest ways you'll affect your organization. Like it or not, you're setting your engineering culture. Take that power seriously. Being a role model doesn't mean you have to become a public figure, be louder than you're comfortable with, or throw your weight around. Many of the best leaders are quiet and thoughtful, influencing through good decisions and effective collaboration (and showing fellow quiet people that there's space for them to lead).

If the idea of being a leader is terrifying, you may need to build up to it. Start small. Maybe compliment someone's success on a public channel, or offer to help onboard a new person. Think of leadership as a skill to build, just like you would learn a new language or technology. The more you practice, the easier it will be.

Be the best engineer and the best colleague that you can be. Do a good job and let others see it. (And help others do the same! We'll discuss how in Chapter 8.) That's what being a role model is.

WHAT DOES IT MEAN TO DO A GOOD JOB AS A SENIOR ENGINEER?

In this chapter, I'm going to spell out the four big attributes that I think you should be modeling. Let's be clear: these are aspirational qualities, skills you should strive to learn and keep learning. I'm explicitly *not* saying that you need to score 100% on each of these attributes to be a "real engineer," or any form of gatekeeping like that; these are ideals. This is how you should *try to be*. We're all works in progress.

One more caveat: the tech industry is awash in advice, most of it subjective. This list is too! Best practices all depend on the situation. There will be edge cases and special circumstances; if my advice contradicts your own judgment, trust yourself. A staff engineer has the good sense to know when the conventional wisdom is wrong.

The four attributes we're going to look at for the rest of the chapter are being competent, being a responsible adult, remembering the goal, and looking ahead.

First up, competence.

Be Competent

As a staff+ engineer, or any senior person, a big part of your role is to take on things that need to be done and to reliably do them well. Competence includes building (and keeping) knowledge and skills, being self-aware, and having high standards.

KNOW THINGS

No matter how good your leadership, you can't be a *technical* leader without the "technical" part. Your big-picture thinking, project execution, credibility, and influence are underpinned by knowledge and experience. A big part of the value proposition of hiring you is your knowledge: you have *seen some things*.

Build experience

Stephanie Van Dyk, staff engineer at Google, draws parallels between the foundational skills needed for her job and those she uses in her longtime hobby as a weaver. "Technical skills come from study and practice," she says. "They aren't inherent. No one is born a skilled weaver; no one is born a skilled computer engineer."

Experience comes through time, exposure, and study—not innate aptitude. It takes work to develop technical skills. You can learn a lot from books, but there's no substitute for working through problems yourself, learning your own techniques for solving them, and seeing what works and what doesn't. Paula Muldoon, a senior software engineer and violinist, describes playing in an orchestra in a way that resonates for me: "You want to get so good at your craft that your focus can be almost entirely on what other people are doing and you don't have to worry about your own execution." Invest time—lots of time—in honing your technical skills so that they become second nature to you.

How much time? The American Society of Civil Engineers publishes its engineering grades (*https://www.asce.org/engineergrades*), and its requirements include number of years of experience:

- Grade V (Typical titles: Senior engineer, program manager): 8+ years
- Grade VI (Typical: Principal engineer, district engineer, engineering manager): 10+ years
- Grade VII (Typical: Director, city engineer, division engineer): 15+ years
- Grade VIII (Typical: Bureau engineer, director of public works): 20+ years

We're less rigorous in software engineering. Job levels vary a lot across companies. Most places don't consider years of experience when allocating grades or titles, but staff engineers typically have at least 10 and principal engineers have at least 15.

Don't rush past your prime learning years. Some organizations encourage their best talent by offering them senior roles, like management or staff engineering, relatively early in their careers. The push for career progression may entice you to accept. But, as Charity Majors warns (*https://oreil.ly/vmIuH*):

> Never, ever accept a managerial role until you are already solidly senior as an engineer. To me this means at least seven years or more writing and shipping code; definitely, absolutely no less than five. It may feel like a compliment when someone offers you the job of manager—hell, take the compliment ☺—but they are not doing you any favors when it comes to your career or your ability to be effective.

It's the same for staff+ roles. Do a whole lot of soul-searching before taking a role that takes you further from the tech. You could be cheating yourself out of your prime fully immersive hands-on experience-building years.

Are You "Technical Enough"?

I've sometimes heard engineers describe other people as "not technical enough," and it's a framing I always push back on. Apart from being a little dismissive, it claims to describe *who someone is* instead of *the skills they have*.[1] It leaves no actionable path to *becoming* "technical enough." If you're inclined to describe someone in this way, be more precise. Do you mean:

- They haven't yet spent enough years doing hands-on technical work?
- They don't yet know the domain you're working in?
- They're missing specific skills? Which skills?

Being competent and knowledgeable doesn't mean you have to know the most about every topic. Sometimes, when you come into a new domain, you will know the least, and that's OK!

Build domain knowledge

Software has an extraordinary number of technology areas, each with its own specialized knowledge and vocabulary. Knowing mobile development, algorithmic computer science, or networking doesn't prepare you for a frontend UX project. Years of experience in fintech won't prepare you for a health care startup. But if you're interested in a new technology area or domain, you can still go try it out. That means that even very experienced engineers can find themselves being beginners.

When you move into a new role, there will inevitably be skills that you don't have yet, or domain knowledge that will be new to you. You may find that you're learning from the junior engineers you work with. (This is a good thing!) Here is where your technical knowledge provides a foundation. While you might not recognize the specifics of the problems in this new domain, your general experience should help you recognize their *shapes*. You should be able to pattern-match what

---

1 There's also sometimes an implication of "doesn't seem nerdy" or "doesn't look like an engineer"; we all should be aware of our implicit bias (*https://oreil.ly/jRVit*). But I'm speaking here to the people who are genuinely trying to help themselves or their coworkers build skills.

you're encountering to something else you recognize, so you're not completely at sea.[2] The broader your scope of experience, the more "hooks" you'll have to hang new knowledge off, and the faster you can learn new things.

When you move into a new technology area or business domain, be deliberate about learning quickly. Learn the technology in whatever way is most effective for you. Get a sense of the appropriate trade-offs; the resource constraints; the common arguments, biases, and in-jokes. Know which technologies are on the market and how they might be used. Read the core literature. Know who the "famous people" in the domain are and what they advocate for. (Twitter is handy for this.) Then take on some projects that will let you build instincts and experience, so you can become as competent in this new area or domain as you were in your last.

Stay up to date

Being a senior engineer means having a growth mindset and a drive to improve. It's embarrassing for everyone when a technical leader insists on a "best practice" that has been debunked for a decade or a technology that everyone else has moved on from. Stay engaged with what's happening in your part of the industry. Even if you aren't deep in the code any more, your spidey-sense should stay sharp for "code smells" or problems waiting to happen. Even if you don't know the latest, hottest tool or practice, know how to find out.

Notice if your role is preventing you from continuing to learn. In particular, be wary of drifting so far from the technology that you're only learning how *your company* operates. While you should *also* learn enough about the business to make good choices, keep yourself anchored in the tech. I'll talk more about learning in Chapter 9.

Show That You're Learning

Junior engineers need to see that lifelong learning is part of being a senior engineer, and that they'll never reach the end of their journey. They also need to see that your skills and knowledge didn't magically

---

2 When I started a new job after 12 years at Google, a company famed for using its own internal technology stack for everything, I relied on this kind of pattern-matching. I did lots of drawing systems on whiteboards and asking, "Is there a thing that looks kind of like this, and you would use it in this situation? Oh, *that's* what Envoy does! OK, got it!"

come to you—it's all learnable. Be open about what you're learning, and show how you're doing it. When you're making a statement that involves obscure knowledge or a logical leap, fill in the gaps for the folks around you: tell them why you came to the conclusion you did, what information you used, and how you came by that information. Be clear that you're learning so it's safe for them to learn too.

BE SELF-AWARE

Competence is built on knowledge and experience, but you also need to be able to apply those abilities. That starts with having the self-awareness to know what you can do, how long it will take, and what you *don't* know. It means being able to say "I've got this" and knowing that you do, in fact, have this. Competence means having well-founded confidence that you'll be able to solve the problem. You don't need to be arrogant, speak in incomprehensible jargon, or show off. True confidence comes from having done the work for long enough that you've learned to trust yourself.

Admit what you know

Some people are brought up to brag about their accomplishments. Others are brought up to minimize them. Whichever you innately are, aim to get to a level where you're confident and honest with yourself about what you know and what you don't. There are going to be some areas where you know a lot and are unusually skilled. Be confident about applying those skills to solve the problems that need them.

Being competent doesn't mean you need to be *the best*. I've sometimes seen tech people be shy about claiming to be an expert, because they can always think of *someone* in the industry who is better than they are. Don't set your bar at "best in the industry." It won't help anyone if you hold back out of modesty. When the skill is needed, don't wait for someone else to say, "Hey, aren't you great at regular expressions" (*https://xkcd.com/208*)? Volunteer that you are—it's not a brag, just a statement of fact.

If you know what skills you bring, then you know where you can step up and help, where you'll be a good mentor, and what you still need to learn.

Admit what you don't know

You won't know everything, and it's vital that you don't pretend you do. If you bluff, you'll lose opportunities to learn—and you may make bad decisions. You'll also waste the opportunity to set an example. Every time you admit you don't know everything and let people see you learning, you show your junior engineers that it's normal to continuously learn.

Admitting ignorance is one of the most important things we can do as tech leads, senior engineers, mentors, managers, and other influencers of team culture. I love asking for an "ELI5," a term that comes from Reddit and means "Explain it like I'm five years old." It's a helpful shortcut to mean "Look, rather than guessing my level of understanding, just spell it out for me. I promise not to be offended if you tell me things I already know." (The social contract here is that you can't get offended if they start with the very basics of the topic.)

We spend a huge amount of our work lives communicating, trying to get a shared model of the world into our brains so that we can collaborate or make decisions. It slows everything down when people in a conversation bluff or disengage a bit because they don't want to be called out on not knowing something. If senior people can admit they don't know things, everyone else will do it too.

Understand your own context

A huge part of self-awareness is understanding that you have a perspective, that your context is not the universal context, and that your opinions and knowledge are specific to *you*. You'll need to escape your echo chamber every time you talk to teams in other areas or explain technical topics to nonengineers. You'll know what information you have that they might not, so you can bridge that gap. (You can read more about building this kind of perspective in Chapter 2.)

It's much harder to explain something simply! It requires more understanding of the topic and more self-awareness about your own context. But it's a real indicator of expertise. If you can explain a topic in plain language, so that nonexperts can hook it onto something they already understand, you really understand it.

HAVE HIGH STANDARDS

Your standards will serve as a model for how other people work. Know what high-quality work looks like and aim for that standard in everything you do, not just the parts you enjoy most. Write the clearest documentation you can. Be the first person to know if your software breaks. There are always trade-offs, of course: sometimes the right move is to slap a solution together with duct tape as quickly as possible. But make that determination based on the problem you're trying to solve, not how interesting the work is to you.

Seek out constructive criticism

Having high standards means making your work as good as it can be. Look for opportunities to put aside your ego and ask someone else to help make your work better. Ask for code review, design review, and peer evaluations. When you've got an idea you love, invite your colleagues to poke holes in it. When you "request comments," don't secretly resent them; each one is an opportunity to make your solution better, so take them seriously even if you don't use them all. Your solutions are not you and they don't define you. Criticism of your work isn't criticism of you. (You'll *give* constructive criticism too, of course. We'll explore that in Chapter 8.)

Own your mistakes

At some point you *will* make a mistake, and it might be a big one. Maybe you reviewed code and didn't notice a bug that cost the company a ton of money. Maybe you wrote that code! Maybe you said something in a meeting that you later found out made someone cry (or quit).

Mistakes are normal.[3] Humans aren't perfect, and mistakes are how we learn. What matters most is how you respond to your mistakes. It's easy to get defensive, deflect blame, or fall to pieces (that someone else needs to pick up). To be competent, you need to own your mistakes. Don't beat yourself up, but don't deny the impact or insist that the mistake wasn't *really* your fault. Admit what happened, then set out to fix it. Communicate quickly and clearly and make sure everyone has the information they need. If there's any risk that someone else will get blamed, clarify that they didn't do anything wrong. If you hurt someone else's feelings, acknowledge the hurt and apologize. (Even if *you* wouldn't have felt bad in the same situation, their feelings are real.)

If you caused an outage or took an expensive wrong path, consider having a retrospective afterward where you talk through what happened, how you recovered, and what you learned. Be open and matter-of-fact about the part you played. It's much easier to understand what happened when nobody's trying to downplay their missteps.

Making a mistake just *stings*. Solving the problem you caused may be the last thing you want to do in that moment. But it's the best thing you can do to retain the goodwill and social capital of your team. If you react well and fix the problem

---

3 If you're thinking, "I don't make mistakes because I'm competent and careful," the gut-punch feeling when you do make one will be so, so much worse.

you caused, you could even end up with *more* esteem from your colleagues. And a leader being open about their mistakes will make it easier for everyone else to do the same: it's a big boost to the team's psychological safety.

Be reliable

My final thought on competence is this: be reliable. One of the biggest compliments I give is, "Alex is going to be in that meeting, so I don't need to go." When I say that, I'm not just saying that any information I have will be represented in the meeting. I'm also saying that I think the right thing will happen. The situation will be managed. I don't need to be there. I'm saying that I find Alex reliable.

A reputation for reliability is like the credibility and social capital we talked about in Chapter 4: it builds up as people see you do the work and get things under control. Be the sort of person who is trusted to get it done well.

Part of reliability is also finishing what you start. Use the techniques in Chapter 6 to make sure you're not blocked or stopping too early. Stick with it even after it gets boring or difficult. And if you stop deliberately because the project isn't the right use of resources, own and communicate that decision. You accepted responsibility for the work, so take it to the finish line.

That brings us to the second attribute senior engineers should strive for: being the responsible person in the room.

Be Responsible

Like it or not, a senior or staff title turns you into an authority figure—and, as the philosopher Uncle Ben once told Spider-Man, with great power comes great responsibility. The more senior you get, the more you have to internalize that nobody else is coming to be the "grown-up in the room." *You* are the "someone" in "someone should do something."

In this section, we'll look at three aspects of responsibility: taking ownership, taking charge, and creating calm.

TAKE OWNERSHIP

Senior people own the whole problem, not just the parts that go as planned. You're not running someone else's project for them: it's *yours* and you don't passively let it sink or swim. When something goes wrong, you don't shrug and decide the work is impossible. You navigate the problem and you're accountable for the result. (Chapter 6 has techniques for getting unblocked that can help here!)

Avoid what John Allspaw calls (*https://oreil.ly/lfRcs*) "Cover Your Ass Engineering" (CYAE):

> *Mature engineers stand up and accept the responsibility given to them. If they find they don't have the requisite authority to be held accountable for their work, they seek out ways to rectify that. An example of CYAE is "It's not my fault. They broke it, they used it wrong. I built it to spec, I can't be held responsible for their mistakes or improper specification."*

Ownership also means using your own good judgment: you don't need to constantly ask for permission or check whether you're doing the right thing. But that doesn't mean you should operate in private. While the classic advice is to seek forgiveness rather than ask permission, Elizabeth Ayer, a product and delivery advisor, offers a more open and predictable approach: "radiating intent" (*https://oreil.ly/fXxG4*), the idea of signaling what you're about to do before you do it. You're giving everyone else context about your actions—and you're creating an opportunity to intervene if you're about to do something dangerous.

Ayer calls out another important advantage of radiating intent: "The 'radiator' keeps responsibility if things go sour. It doesn't transfer the blame the way seeking permission does." That's key to ownership too.

Make decisions

Professional engineers in some disciplines are responsible for putting their seal on documents: an engineer might sign off on the structural integrity of a building, for example. By doing so, they're attesting that the document is structurally safe and taking on legal liability for any mistakes they've made. They're personally on the hook for that decision.

While software engineers don't currently have this kind of professional responsibility, as technical leaders, we must be prepared to make the final call and own the outcome. In particular, when a decision is needed, avoid staying on the fence: weigh the options, choose decisively, and explain your reasoning. Be honest with yourself as you consider the trade-offs: you should be able to vote against your own preferences when you know it's the best move.

Owning decisions includes accepting that you might be wrong. Make the cost of a wrong decision as low as possible, and if it turns out that you *are* wrong, own that, too.

Ask "obvious" questions

One of the best things about being senior is that you can ask questions that are so obvious, nobody else is willing to ask them. Here are a few examples:

- It sounds like you're planning to run a mission-critical microservice in a team with only two engineers. How do you intend to handle on-call for it?
- I assume you've evaluated what it would take to move off this old system instead of working so hard to keep it alive?
- What will happen if users start to depend on that incrementing field in your API that you're telling them to ignore?
- You've run this odd-sounding proposal by security, right?
- What would it take to support this use case that we keep asking people not to do on our platform?

As a leader, you have a responsibility to make the implicit explicit. It's not fair, but if a junior person asks these questions, the team may sigh and say, yes, *obviously* we thought of that. If an expert asks, team members learn that they should include explicit answers to these questions in their design documentation. (Or they genuinely consider the question for the first time!)

Don't delegate through neglect

A few years ago I wrote a conference talk that went a bit viral. OK, we're not talking *otters holding hands* viral, but it swept across tech Twitter, hit the front page of Hacker News, that sort of thing. The talk was about the leadership and administrative tasks that aren't on anyone's job ladder but are needed to make a team successful: all the unblocking, onboarding, reminders, mentoring, and scheduling. I called this kind of work "glue work" (*https://noidea.dog/glue*).

Why did the talk hit such a nerve? Because although projects can't succeed without it, this kind of work is rarely rewarded or allocated fairly. It falls to whomever on the team can't look away from the problem, often a junior person with a strong sense of ownership.

The problem is, when junior people do too much administrative or leadership work and not enough technical work, they're spending their prime technical learning years in a way that doesn't teach them technical skills. That can stunt their careers in the long run. But, often, leaders don't step in: the glue work is needed for the project to succeed, and they're just glad it's getting done.

If glue work is needed for your organization or your project, recognize it and understand who is doing it. Be aware that managers, promotion committees, and future employers might consider this work to be *leadership* when a staff engineer does it, but dismiss it when a more junior engineer does. So take ownership and do a lot of the work that's not anybody's job but that furthers your goals. Redirect your junior colleagues toward tasks that will develop their careers instead.

TAKE CHARGE

Gently redirecting your colleagues towards more valuable work is an example of what I'm going to talk about next: taking charge of the situation. Note that *taking charge* doesn't necessarily mean you have prior authority. It means that you see a gap and you're stepping up to fill that gap.

Step up in an emergency

Being able to take control of a mess is a key aspect of technical leadership. If security detects a breach, a database gets dropped, or a meteor hits US-East-1, there are likely to be many responders. Unless they're working together, the ensuing chaos can make the problem worse. Everything goes better if someone is coordinating.

Unfortunately, coordinating only works if everyone *knows* you're coordinating. If you don't take charge explicitly, you'll just be one more person making noise. Ideally, you'll have emergency plans in place before the disaster hits, so that the role of the coordinator is well understood: the classic Incident Command System is a popular choice.[4] If not, you're going to have to find a way to announce that you're coordinating and set expectations about what you're going to do. Then make sure your coordination is valuable. A few ways to do this are to take clear notes, make sure that everyone involved in the emergency has the same context, and ask everyone to *radiate intent* about what they're doing and when.

---

4 The Incident Command System (*https://oreil.ly/DUxaG*) was introduced by fire departments in the '60s and is now used by most emergency services in the United States to coordinate disaster response. One of the roles it defines is the incident commander, someone whose job is not to fight the fire, but to coordinate and take command. It works well for the kinds of software outages that are chaotic or that cross multiple teams.

Ask for more information when everyone is confused

Earlier in this chapter, I talked about admitting what you don't know and asking obvious questions. During an emergency, you'll often need to do both at once. When teams are sharing information to resolve the issue, they often don't all have the context to interpret it. Objective facts like "the FooService has 1% 401 errors" aren't helpful to anyone who doesn't know what's typical for the service. Is that bad? Is there a theory for what's happening? How does FooService fit into this outage?

Someone needs to be brave enough to say, "I don't know what to do with the information you just gave me!" Take charge and ask. Tech can be fraught with egos and insecurity, and it's sometimes scary (or legitimately risky!) for junior people to admit that they don't know something. It's safer for senior people to ask.

Feigned Surprise

Feigned surprise used to be a standard part of conversation for sysadmins and software engineers: "You've *seriously* never used Linux?" It was part of the BOFH (*https://en.wikipedia.org/wiki/Bastard_Operator_From_Hell*) toolbox, designed to undermine those who were still learning (the noobs and the lusers) and let the more experienced tech folks feel superior.

How do you ask questions in that environment? Mostly, you don't. You make a poker face, try to keep up, and hope the topic changes to something you know. When you absolutely can't avoid it, you ask for help in private. It takes much longer to learn anything.

But then the Recurse Center (then called Hacker School) called out the phenomenon in its social rules (*https://oreil.ly/WLkOr*). Naming the behavior gave people power over it: it was something they could recognize and ask others not to do. The Recurse Center built in a mechanism to keep the rules low-stakes too: "The social rules are lightweight. You should not be afraid of breaking a social rule. These are things that everyone does, and breaking one doesn't make you a bad person. If someone says, 'Hey, you just feigned surprise,' don't worry. Just apologize, reflect for a second, and move on."

Don't feign surprise, but go even further. Every time someone apologizes for asking a basic question and there's a chorus of reassurance from others who insist that it's actually a *good* question, culture is built. That's an environment where it's easy to learn.

Drive meetings

Meetings are another place where it really helps to have someone step up and take charge. If the group is passive, distracted, or inclined to turn a work meeting into a social conversation, any one of the attendees can (in theory) say, "OK, let's get started on our agenda." But most meeting attendees are hesitant to play that role. Step up when it's needed. Make sure there *is* an agenda: collect items to discuss at the start of the meeting, or set the example of sending around the agenda in advance. Remember what you're hoping to get out of the meeting, and drag it back to that topic if it goes too far astray.

If the meeting doesn't have notes, was it really worth getting together? Meeting notes are a great example of glue work. If a junior person is taking notes, they're unable to participate, and it's considered low-status administrative work. If a senior person takes notes, they're making sure the meeting is effective, and everyone's very impressed!

Meeting notes are a great lever for making progress on your projects, so don't hesitate to volunteer to take them. You can record the facts you think are most important, document decisions made, and be the first to frame the decision. Then you can invite everyone to confirm what you wrote. As a moderator, if you need to give everyone a moment to think and reflect, you can also say, "Wait a moment, I need to catch up with the notes." They're a useful flow control for the meeting.

If you see something, say something

Another common situation where you might need to take charge is an awkward one: when someone just said something disrespectful or offensive in a public channel. Other people in the room might want to say something but feel like they lack the social capital. Use your position as a leader and speak up.

Like many engineers, I find these situations uncomfortable, so I asked Sarah Milstein, engineering VP at Daily, for advice. She always seems fearless when confronted with "advanced humaning" problems, so I was disappointed to learn that speaking up isn't magically easier when you're a manager. The adrenaline awfulness, she told me, doesn't go away. You just accept the discomfort. Go in knowing that it's going to be awkward, but that it will be better to have said something than not. You don't have to say the perfect thing—there often *isn't* a perfect thing—but you do need to speak up.

While the conventional wisdom for feedback is to praise in public and criticize in private, this is a time when it's vital that you say something public: if it looks like the original message wasn't addressed, it can create an environment where that kind of message is seen to be acceptable. If someone is attacked in front of a group, you need to support them in front of the same group. If nobody addresses the problem, your group dynamics will become weird and uncomfortable.

It's best if you can deal with this kind of situation quickly, but it's OK to return to it a little later: "Look, I feel this hanging in the air and I wish I'd addressed it at the time, but I want to go back to it." By addressing it, you can "give the energy a place to travel," Milstein says. She adds, "Almost always, somebody thanks me later for having spoken up in a hard situation."

Here are some more techniques from Milstein:

- Describe the culture that you're aiming to build, and use that as a reference. For example, "You all know that respect for each other is a big value here. It's part of how we get things done. That message violated those norms."

- Give the person a path to being on the same side as you. For example, if they made a hurtful joke about a tense news story, show that you understand why someone would joke just then. "I get using humor in hard situations, but let's be mindful that people in this meeting might be affected by what's going on."

- If it's a private conversation, you can appeal to the person's values: "I know you really care about fairness, so I want to flag something you said that you might not have realized the implications of."

Finally, this isn't a thing that should end with you. While you have the power and responsibility to address culture issues, this situation is a behavior problem too. You can also tell a relevant manager so that, if there's a pattern, they can help address it. That's their role, not yours.

CREATE CALM

The final factor in being the responsible grown-up in the room: stay calm. Tired, stressed people often disagree about the right way to proceed. If you can stay calm and constructive and avoid casting blame, other people will too.

Defuse, don't amplify

If you're dealing with a big problem, try to make it smaller. If you're dealing with a small problem, *keep it small*. When someone brings you a fraught situation, stay calm. Ask questions. Understand why they're telling you. Do they just need to vent? Are they hoping you'll take action? Be curious, even about topics you think you understand. If there's a problem, acknowledge it. Even just seeing that you have the same information and don't seem to be panicking can be enough to reduce a colleague's anxiety.

Even if you can see something you can do, don't react reflexively. A senior person making a fuss can blow up a minor thing into a big, loud issue, so be certain that you have all of the facts and that you're genuinely helping by joining in. If your actions will amplify rather than calming the situation down, consider staying out of it. Also, remember the warning at the start of Chapter 6: make sure this side quest is the right use of your time.

Finally, be cautious about where you share your anxieties or frustrations. While you can acknowledge that there are problems, don't let your worries about them spill out on more junior people: it's not fair to ask them to carry your concerns, and you're amplifying the problem if you upset them. That's not to say you have to keep your worries to yourself: you can vent to your manager, close peers, or the project sounding board you chose in Chapter 5. But be clear about whether you're complaining socially or you want the other person to do something. Be especially clear in one-on-ones with other leaders about whether you're asking for action, unpacking something for yourself, or sharing context. *They* might reflexively react and amplify something that you'd just intended to blow off steam about.

Avoid blame

I still remember one of the first mistakes I made in production. While updating a customer record, I'd somehow deleted their entire account. I was 22, new to the team (and the industry), and absolutely petrified that taking the blame would mean the end of my short career. My coworker Tim cleaned up my mess and I'll never forget his reaction: "It's always interesting to see how new people handle their first screw-up. We've all been there." It was such a relief! Of course I was still upset, but the sick feeling in my stomach was gone. If every one of my coworkers had survived their first mistake, I would too. In the middle of the annoying task of recovering the customer data, Tim took the time to be kind.

A big outage is an expensive training course, and if you're paying the cost, you'd better all learn something! If someone made a mistake or discovered an edge case by breaking something, create an environment where everyone will feel safe talking through the event. If you're curious and avoid blame, you'll find it easier to ask question like:

- Exactly what happened?
- What factors led them down this path?
- Was there information they didn't have, but could have had?
- Where did their mental models diverge from reality?

Be consistent

Have you ever had a leader who is a complete wild card? You don't know how to prepare for any meeting with them. One day they only care about high-level project delivery dates; the next, they're asking you to justify the tiniest technical decisions. They tell you that one goal is the most important business need and then, just as you've finished adjusting your project plans, they prioritize something else. It's chaotic. You can't know where you stand.

Don't be that leader. Instead, create a sense of safety and calm by being consistent and predictable. Your colleagues should know what they can expect if they ask you to help with something. During times of change or difficulty, the way you show up and express yourself at work can reassure your colleagues: yes, change can be scary, but they can rely on you to be steady while you all work through it.

It's harder to be consistent when you're stressed out or working beyond your capacity, so being consistent means taking care of yourself. Remember Figure 4-3: leave a little space in your life for unplanned events. Work in a way that is sustainable for you. That means taking time off, getting enough rest, and doing the things outside work that make you happy. Remember, you're modeling sustainable work for your colleagues too.

Remember the Goal

On to the third attribute of being a role model senior engineer: remembering what the heck you're all doing here. It's not just technology! There's a broader context: a business that's trying to achieve something, a mission you're setting out on. In this section, we'll look at bringing business context (and budget context) to your decisions and solving the entirety of the bigger problems your users care about, not just your team's tasks. And we'll think about achieving the goals as a team, not as individuals.

REMEMBER THERE'S A BUSINESS

As a senior engineer, you have a responsibility to the future as well as the present. You will always be responsible for creating software that stands up under stress. But you're working for a business (or a nonprofit, government agency, or other organization) that has goals. The software is the means to those ends, not an end in itself.

Adapt to the situation

I took part in a hackathon once for a volunteer event, and I remember the team lead looking at my code and saying, "Wow, tests? That's, uh, nice." He was being polite, but it was clear that tests were unusual—and not particularly welcome. Speed was much more important than accuracy, and the code was going to be thrown away later. As far as he was concerned, I'd wasted my time.

Sometimes a faster solution is better; sometimes a more stable one is. If time to market is vital to your business's survival, getting a shoddy first version out the door might be more important than having beautiful code and architecture.[5] Similarly, if you're shipping software to accompany a holiday promotion or a major sporting event, an imperfect solution is much better than a late one.

Priorities sometimes change during an outage, too. Maybe you've got a rule that you always do a clean rolling restart of your service so that you only have a couple of instances offline at a time. When everything is broken, though, your usual principles might go right out the window: sometimes the fastest thing to do is to turn the whole thing off and then back on again. A midlevel engineer might strive for the platonic ideal of a clean, technically elegant fix, but their

---

5 But don't help a business to survive at the cost of releasing software that endangers, exploits, or hurts other people: the company has bought your time and energy, but not your moral compass. We'll look more at values in Chapter 9.

senior mentor will teach them that this is when you get the system back online first and clean up later.

As the business changes, your priorities will change. *Be OK with that.* It's inevitable. Growth, acquisition, new markets, or a change in fortunes will mean that your goals may get thrown out as the company pursues a new direction or even a new culture. If you don't like that or it doesn't fit your values, you might no longer be in the right place. But if you just resent change, you'll spend your time being unhappy. Expect it, and you'll embrace it as a new challenge instead.

Be aware that there's a budget

Your high engineering standards will always be in tension with the amount of money the business is willing to spend on good engineering. That tension doesn't mean you should drop your principles and start advocating for shoddy software, but do keep the idea of a budget in the back of your mind. Remember that other people are limited in what they can spend on headcount, vendor tooling, and so on.

Don't obsess about the budget: it's easy to get frozen in indecision, trying to decide if one project or another is really worth the cost. But have a feeling for what kinds of expenditure, savings, or new revenue are considered big. Understand how your company makes money and know whether you're in good times or lean times. Bear those facts in mind when you're deciding what to suggest your organization spends time on.

Spend resources mindfully

"Growing up" in Google during a time of plenty, it took me a decade longer than it should have to realize that headcount is finite and staffing a project has an opportunity cost. Part of your technical judgment is "spending" that finite headcount wisely.

You'll probably have a ton of ideas about places you can innovate, invent something new, or make one of your systems a little better. Make sure you're choosing work that your business actually needs. Your team has finite time and energy; is this the right way to spend it? Take Dan McKinley's advice (*http:// boringtechnology.club*) too and be judicious about where you spend your "innovation tokens": that is, your company's "limited capacity to do something creative, or weird, or hard." If you can only invest in a few places, is this the right place?

Build the most useful thing, not the thing that would be most fun to build. And when it's time to stop polishing something and declare that it's good enough, stop.

REMEMBER THERE'S A USER

I remember once sitting with a vendor in a huge cafeteria with hundreds of coworkers. My colleague Mitch and I listened as the vendor explained how a feature we'd been asking for was now ready for us. But I'd tried the feature and it didn't work. We argued back and forth until I pulled out my laptop and showed him.

"Oh, I get it now," the vendor said. "You're using Chrome. It works on Firefox and Internet Explorer." (Yeah, this was a while ago.) "But don't worry, not many people use Chrome."

"Look around you," Mitch replied, gesturing around the cafeteria. "See all of these people? Everyone in this room uses Chrome."

I've seen too many teams create a feature for a set of fictional, perfect users who don't exist.[6] Know who uses your software and know how they use it. Make sure they can use the thing you're creating for them, and that they *want to.*

Part of getting this right is the classic solution: write it down! Be clear about the exact requirements you're creating for, and share those requirements broadly. Get the proposed API reviewed before you start the code. Show a mockup of the user interface before you start creating it. Check in frequently and show updates. Once again, avoid CYAE: whether you built it to spec or not, if you didn't make your user happy, you built the wrong thing.

REMEMBER THERE'S A TEAM

The final thought in focusing on the mission: remember you're not doing this work alone. While you may be the best coder on the team, the most experienced engineer, or the fastest problem solver, that doesn't mean you should jump on all of the problems. You're working as part of a team, not a collection of competing individuals. Don't become a single point of failure where the team can't get anything done when you're not available. It's not sustainable. It hides problems.

Just like I advised you to be self-aware, be aware of the capabilities of your team. If you can reach your goal by empowering someone else to do better work, that's just as much a victory as if you solve it yourself. Consider your impact to be what wouldn't have happened without you, not just what you personally did.

---

6 Perfectly spherical users, as some friends say.

Look Ahead

While there are, as you've seen, times when your first priority will be to get something to market quickly, most of the time you're planning for a longer time horizon. The code and architecture you work on are likely to still be in use in 5 or 10 years. The interconnected software systems that make up your production environment may last much longer, and each component will influence the ones that follow.[7] As Titus Winters writes in *Software Engineering at Google* (O'Reilly), "Software engineering is programming integrated over time." Expect the impact of your software to stick around.

Your organization, codebase, and production environment probably existed before you joined them. They'll probably exist after you move on. Don't optimize for *now* at the cost of future velocity or engineering ability. It's OK to plant some seeds that you won't personally see grow.

Here are a few ways you should be thinking beyond the current moment.

ANTICIPATE WHAT YOU'LL WISH YOU'D DONE

Remember our question from in Chapter 3: "What will Future You wish Present You had done?" When you're making plans or doing work, consider your future self and your future team to be stakeholders: after all, they'll have to deal with whatever decisions you make now.

Telegraph what's coming

Be clear about what your broad direction is, even if you don't know the details yet. Here's an example: teams sometimes avoid announcing deprecation dates for old systems, because they're not quite ready to begin the major migration to the new system. But you can announce the *intention* to deprecate it. If everyone knows a migration will begin in a year or two, new projects will know not to invest in it. Some teams may find themselves with free time and move to the new system without you even asking them. A small amount of work now will set people's expectations, save their time, and make your future deprecation project a little easier.

---

7 Think of it as a Ship of Theseus (*https://oreil.ly/nBaaK*): every individual component may get replaced over the years, but the fundamental system continues. It's all metaphysical architecture.

Tidy up

Have you ever had to work in a tool shed where the last person didn't clean up after themselves? It's horrible. You grab the drill and the battery's out of power. The safety goggles aren't in their case; you search through three boxes before finding them with the sander. The floor is covered in detritus. There is no flow state in an environment like that. Everything takes three times as long as it should.

Now think about what it's like when every tool you want is at arm's reach. Your workflow just *works*. So take the time to leave your production environment, codebase, or documentation so that it *just works* for whoever comes along next. Write tests that will let you refactor your code without breaking things. Follow your style guide so that the people who copy your approach will also be following your style guide. Leave no traps, like dangerous scripts that everyone needs to remember not to run or configurations that are changed locally but not updated in source control. Make it safe to move around.

Keep your tools sharp

Don't *just* tidy up: continually invest in making your environment better. If you can move quickly and safely, you'll spend less time on repetitive work and you'll be able to do more. Increasing your velocity increases your reliability, too: every minute you shave off your time to detect a problem or deploy a fix is a minute you've taken off every outage.

Look for optimizations that will let you build, deploy, and release more quickly: smaller builds, intuitive tooling, fixing or deleting flaky tests, repeatable processes, automation everywhere. Be judicious about where you invest: building tooling, platforms, or processes takes time, so choose the optimizations that will genuinely make a difference.

Create institutional memory

Every time someone leaves your company, you lose institutional knowledge. If you're lucky, you have some old-timers storing history in their brains. But eventually, inevitably, you'll have complete staff turnover.[8] When an old system breaks, there'll be nobody left to say "Oh, yes, I remember when we ran into this before. Here's what we did last time."

---

8 Another Ship of Theseus! The people have all changed but the organization remains.

My ex-colleague John Reese, at the time a principal engineer at Google, often also took the role of systems historian: he curated a record of how the site reliability organization had evolved and how running software in production had changed over the years. To create institutional memory, he wrote in-depth articles about the parts of the ecosystem he knew best, then interviewed others to uncover the past, documenting formative systems and practices. Although he's moved on from Google now, that history lives on with a new set of curators.

While most organizations don't have someone deliberately writing down their history (though maybe we should!), you can send information into the future by writing things down. This includes decision records that explain what you were thinking, systems diagrams that include the obvious things that "everyone knows," and code comments that include context on what's going on. However you create the history, include searchable keywords so that future people have some chance of understanding what you did and why—and think about what you know that future people might not.[9]

EXPECT FAILURE

My all-time favorite incident retrospective is the one Fran Garcia wrote (*https:// oreil.ly/zsPgE*) about his then-employer, Hosted Graphite, being taken down by an AWS outage. The reason I love this one is that Hosted Graphite didn't *use* AWS, so the team was quite surprised at being affected by its outage.[10] They had no way of predicting it.

How many unpredictable failures like that lurk in your systems? Assume it's a *lot*. The network will fail (*https://oreil.ly/OuP1u*), the hardware will fail, the people will have an off day. There will always be bugs. Odd interactions between parts of the system you haven't even thought about will cause problems.

You can't predict everything that will go wrong, but you can predict that *something* will go wrong. Plan for what you'll do when it does. Build the

---

[9] Be inspired by the Sandia National Laboratories report (*https://oreil.ly/PWTxV*) on creating pictographic information to deter future humans from interfering with nuclear waste repositories in 10,000 years when current languages will be long gone. You don't need to think quite that far ahead, but imagine if the systems you work with are still around in 10 years: what will people need to know? How can they accidentally hurt themselves?

[10] In case you're curious: the outage meant that a lot of Hosted Graphite's users became slow all at once and their usually short-lived connections stayed open, increasing the number of connections until they reached a limit in the load balancer and prevented anyone else from connecting. The write-up is a good time.

expectation of failure into your products: test the error paths as thoroughly as the success paths, and make the product do something sensible and user-friendly when it doesn't get the kind of response it expects. Make sure you'll find out when your systems aren't behaving, and have a plan for how you'll respond to it.

Plan in advance for major incidents by adding some conventions around how you work together during an emergency: introduce the incident command system I mentioned earlier, for example, and practice the response before you need it. Your disaster plans will invariably have something go wrong, so simulate disaster with chaos engineering (*https://oreil.ly/NWys8*) tooling or controlled outages. Drills, game days, or tabletop exercises can let you uncover which parts of your response won't work. And of course, if you haven't tested restoring your backups, assume you don't have any backups.

OPTIMIZE FOR MAINTENANCE, NOT CREATION

Software is created once, but it will need to be maintained for years. If you've got a binary running in production, it will need monitoring, logging, business continuity, scaling, and so on. Even if you intend to never touch the code again, the technical or regulatory ecosystem may force you to care: think of all the old systems that needed to be updated for Y2K, to support IPv6 or HTTPS, or for compliance concerns like SOX, GDPR, or HIPAA. Those won't be our last disruptive changes.[11]

Software gets maintained for much longer than it takes to create it, so don't build code that's hard to maintain. Here are some ways you can help Future You and your future team.

Make it understandable

At the moment you create new code or design a new system, you understand it well. Probably the people on your team also have a strong mental model of how it works. Expect that knowledge to decay a little every day. The system will never again be as well understood as it is on the day it's created. If it's hard to understand then, good luck in two years, when something breaks and you're trying to load that mental model back into your brain.

You have two choices to let future people understand your system.

One option is to focus on education and hands-on experience. You can run continual classes about the system, making sure that everyone who might have to

---

11 2038 is coming (*https://oreil.ly/SOdWl*)!

work on it in future is fully trained and has logged enough hours to handle any problems that might arise.

The other option is to make it as easy as possible for people to understand the system when they need it. That means writing documentation with that future person as the main audience: a clear, short introduction; at least one big, simple picture (use arrows to show which direction data moves); links to anything they might wonder about. Then expose the system's inner workings as clearly as possible. Make it possible to see what it's doing, through tooling, tracing, or useful status messages. Make your systems *observable*: easy to inspect, analyze, and debug. And keep them simple, which I'll talk about next.

Keep it simple

There's a Martin Fowler quote that I love: "Any fool can write code that a computer can understand. Good programmers write code that humans can understand."[12] Senior engineers sometimes think they can demonstrate their prowess with the flashiest, most complicated solutions. But it's easier to make something complicated. It's much harder to make it simple!

How can you make something simple? Spend more time on it. When you think of a solution to the problem you're working on, treat it as "just the first." Spend at least the same amount of time on another solution. Now that you understand it better, see if you can make it simpler: fewer lines of code, fewer branches, fewer teams, fewer hours of maintenance, fewer running binaries, fewer files touched.[13] The longer the system is intended to last, the longer you should spend trying to make it as simple as you can. Make it easy to build mental models of the system or the code.

Beware of organizations that seem to reward complexity. Ryan Harter, a staff data scientist, has written about (*https://oreil.ly/Su2IS*) how he's seen people create complicated solutions to prove that they're doing hard work. "I've seen folks slip machine learning into places it doesn't belong to get a flashy launch." He cautions, "Really, what we should want are simple solutions to complex problems. The complexity of our work is a cost to bear, not something to maximize!"

When you're dealing with inherently complex problems, make a deliberate decision about where in the system you're going to put the complexity: that one

---

12 *Refactoring: Improving the Design of Existing Code* by Martin Fowler et al. (Addison-Wesley).

13 If it's so few lines of code that it's getting obfuscated and complicated again, you went too far. We're aiming for understandability, not stunt programming.

terrifying module with the inscrutable business logic or performance optimizations. Make it so that someone looking at the entire system can treat that component as a magic black box and reason about everything else, so that there's a single place to go to when it's time to understand and modify the complex part.

Build to decommission

Someday your system will be turned off. How hard is that going to be for the people working on it then? Will they have to dig deep into the logic of other systems, unwinding tendrils that touch business logic and tracing through code to understand what data they're accessing? Or will there be a clean interface and a simple cutover?

Your architecture will evolve, and your components will settle into the middle. While it might be faster now for you to just wire in the new system, library, or framework, think about what will happen afterward. Will it be possible to replace it later without demolishing whatever other people have built on top of it?

Imagine knowing that *you personally* will need to decommission this component in 10 years. Future You won't be any less busy than Present You, so what can you do to help them out? Might you add a clean interface, make it easy to see which clients are still using a server, or design in a way that keeps a little distance between two systems that are being integrated? If you set out from the start to build a component that's easy to decommission, you'll have the side effect of building something modular and easy to maintain.

CREATE FUTURE LEADERS

Building up your team is an important part of future planning. It often will be easier and faster for you to solve problems or lead projects than for others to do it, but that doesn't mean you should take over. Your junior engineers are future senior engineers. Give them the space to learn, and opportunities to do hands-on work and solve increasingly difficult problems. Chapter 8 will have a lot more about how to continually raise their skill levels.

I'll leave you with one more quote from John Allspaw's "On Being a Senior Engineer" (*https://oreil.ly/aANg3*):

> *The degree to which other people want to work with you is a direct indication of how successful you'll be in your career as an engineer. Be the engineer that everyone wants to work with.*

If you take nothing else away from this chapter, take that last sentence: *the metric for success is whether other people want to work with you.* If they don't, reevaluate your approach.

To Recap

- Your words and actions carry more weight now. Be deliberate.
- Invest the time to build knowledge and expertise. Competence comes from experience.
- Be self-aware about what you know and what you don't.
- Strive to be consistent, reliable, and trustworthy.
- Get comfortable taking charge when nobody else is, including during a crisis or an ambiguous project.
- When someone needs to say something, say something.
- Create calm. Make problems smaller, not bigger.
- Be aware of your business, budgets, user needs, and the capabilities of your team.
- Help your future self by planning ahead and keeping your tools sharp.
- Write things down, even when they're "obvious."
- Expect failure and be ready for it.
- Design software that's easy to decommission.
- The metric for success is whether other people want to work with you.

8

Good Influence at Scale

How do you raise the skills of the people around you? As a staff engineer, part of your job is to enable your colleagues to do better work, to create better solutions, to be better engineers. We already started on this journey last chapter with the idea of being a *role model engineer*: doing the best engineering work you can and letting others see it. That's what we usually mean when we say someone is "a good influence." They behave in the way we'd like others to behave.

But now we'll go further and look at more active ways you can use your good influence to improve other people's skills and your organization's engineering culture.

Good Influence

When you work with someone who is missing skills or has lower standards than you, don't get frustrated: take the time to bring them up a level.

Why is it so important to help other engineers do better work? First, good engineering inevitably goes beyond yourself. If your colleagues do better work, you can do better work too. While some engineers are extraordinary solo artists, even the most powerful virtuoso will meet some problems that are too big to solve alone. If you can help your colleagues become better engineers, you'll be working with more competent people, which means your own work will be easier (and less annoying). Better engineers means better software, which means better business outcomes.

The second reason is that the industry keeps changing. Even if your engineering organization is on the cutting edge right now, at some point there'll be a new game-changing architecture, tool, or process that you want everyone to adopt. Getting the teams you work with to use it will be an exercise in frustration unless you know how to influence and teach new skills.

The last reason is that it's just the right thing to do. As a senior person, you have outsize influence on how well your organization creates software, and even on how our industry behaves and evolves. In the same way that you take pride in improving your code quality, reliability, and usability, you can take pride in your high standards. If you teach your midlevel colleagues to be fantastic engineers, think of the midlevel engineers *they'll* be teaching in 10 years. You're sending high standards into the future.

SCALING YOUR GOOD INFLUENCE

For most of us, leadership through influence starts with individual relationships: reviewing someone's code, hosting an intern, or mentoring a new grad. You might progress to leading small teams, probably having one-on-one meetings with each person on the team. As your *scope* grows, though, it becomes harder to have enough influence purely through individual interactions. There just aren't enough hours in the day. As Bryan Liles, a principal engineer at VMware, describes it (*https://oreil.ly/ScVHa*): "My job at VMware is to be able to influence 14,000 engineers...I'm trying to think 'What can I do to make 14,000 engineers better?'"

Depending on your seniority, your scope, and your aspirations, you might be aiming to influence far fewer people (or many more!). In this chapter, we're going to look at good influence at the micro and macro levels: from improving the skills of your coworkers, team, or group to changing the trajectory of your whole organization or even the entire industry. I'll describe three tiers of influence (see Figure 8-1):

Individual

You're working in a way that grows another person's skills.

Group

You're scaling your influence by bringing new skills or a change of approach to multiple people at once.

Catalyst

The change you make goes beyond your direct influence. You're setting up frameworks or community structures that let your positive influence continue even after you step away.

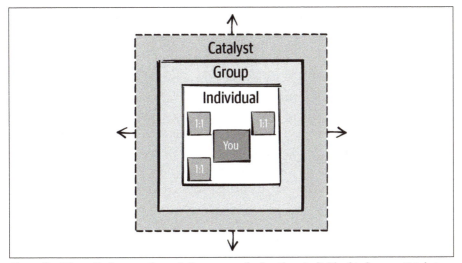

Figure 8-1. Tiers of influence. Group influence goes further than individual influence. Catalyst influence keeps going even when you stop investing more effort.

What forms can influence take? I'll describe four mechanisms for bringing your colleagues up a level, and give examples at each tier:

Advice

We'll start with giving advice, both solicited and unsolicited. At the individual level, this can mean mentoring, peer feedback, or just answering questions. You can scale your advice to the group level through writing and presenting, and get to the catalyst level by making it easy for your colleagues to advise each other.

Teaching

We'll look at deliberately teaching skills to individuals through training, pairing, shadowing, or coaching. Then we'll scale to groups using onboarding materials, codelabs, classes, and workshops. The catalyst level is teaching other people to teach, setting curricula, and influencing the topics that everyone is exposed to.

Guardrails

We'll explore how to give people guardrails so they can work safely. For individuals, I'll talk about code review, design review, and how to be someone's project guardrail. For groups, we'll look at some of the processes, policies, and robots that can keep us on track. Finally, at the catalyst level, we'll explore the ultimate guardrail: culture change.

Opportunities

Finally, we'll look at helping people grow by matching them with opportunities that will help them learn. For individuals, I'll talk about delegation, sponsorship, and highlighting good work. For groups, the biggest opportunity you can give your team might be stepping back, making space, and sharing the spotlight. And it becomes a catalyst when the entry-level and midlevel folks you sponsored become dynamic, skilled, problem-solving senior people who take on challenges and create new opportunities that you never imagined.

Table 8-1 shows these three tiers and four mechanisms, with some examples.

Table 8-1. Some examples of scaling advice, teaching, guardrails, and opportunity across the organization and beyond.

	Individual	Group	Catalyst
Advice	Mentoring, sharing knowledge, feedback	Tech talks, documentation, articles	Mentorship program, tech talk events
Teaching	Code review, design review, coaching, pairing, shadowing	Classes, codelabs	Onboarding curriculum, teaching people to teach
Guardrails	Code review, change review, design review	Processes, linters, style guides	Frameworks, culture change
Opportunity	Delegating, sponsorship, cheerleading, ongoing support	Sharing the spotlight, empowering your team	Creating a culture of opportunity, watching with pride as your superstar junior colleagues change the world

Think of these tiers and mechanisms as a list of options available to you, not a checklist to try to complete. Play to your strengths and do the ones that you enjoy, find easy, or want to get better at. And although they're framed as a hierarchy, don't skip past the "smaller" ones. Leveling up your colleagues through code review or sponsorship will have a ripple effect across the company, and even the most world-changing technologists still mentor people they believe are worth investing in.

Similarly, don't get too focused on chasing the *catalytic* types of influence in the right-hand column of Table 8-1. Having too many programs and frameworks can be overwhelming for your organization, and you will often have more impact by taking part in an existing initiative than by setting up something new. If there's already an onboarding curriculum, for example, teaching a class in it will usually be more valuable than setting up a separate education initiative. Navigating a single team through a difficult design can often be much more important than tinkering with the RFC process. Do the individual and group work first, and only go broader if the need and value is very clear.

All that said, let's start with advice.

Advice

"Free advice," the maxim goes, "is worth exactly what you paid for it." And it's true that advice is *noisy*: there's bad signal mixed in with good, and it tends to not be tailored to the person who's receiving it. However, everyone needs advice sometimes, and giving it is one of the ways you can pass on your experience. If you've had successes and made mistakes, let other people learn from them too.

Who Asked You?

Before you offer advice, think about whether it's welcome. Solicited advice is when someone else asks for something: a recommendation, feedback on their work, help deciding what to do. When it's unsolicited, they haven't asked.

Before you offer your thoughts, think about whether the other person is asking for them. Think too about whether you even have enough context to tell them something that's both helpful and nonobvious. If you're not sure whether your advice will be welcome, *ask them*.

There are times when unsolicited advice is valuable. "Your slides were amazing, but you talk to your shoes when you present and it's hard to understand you" is difficult but probably kind unsolicited advice.[1] But think about your role and your relationship with the person. Some advice should come from friends, not strangers. And, again, don't just launch in. Ask "Can I offer some advice?" and get their permission.

---

1 "Watch out, there's a bear behind you" is also acceptable unsolicited advice.

Here are some other places where unsolicited advice might be helpful:

- When you think the person is in a situation they can't see out of. "Hey, I know this isn't the question you're asking, but that job situation you described doesn't seem healthy. You deserve better."

- When you have key information the other person doesn't have. "I heard you say you're going to start using the Foo platform; just want to make sure you know it's going to be turned down next year."

If you're itching to give unsolicited advice on a topic nobody is asking you about, consider writing a blog post or tweeting about it instead.[2]

INDIVIDUAL ADVICE

There are a few ways that you might give someone individual advice on a situation they're in. These include being a mentor figure to them, answering their questions, commenting on work they did, or giving peer feedback at performance review time.

Mentorship

Mentoring is often an engineer's first experience of leadership. In companies with formal mentorship programs, you might be assigned to help a new person get oriented. Mentorship can happen organically too: sit next to someone, introduce yourself, and answer their questions, and you might find yourself with an informal mentee.

By sharing your perspective and what you've learned, you can accelerate other people's learning and save them from making unnecessary mistakes. You can tell them stories about similar situations you've been in, what you did, and what the outcomes were. Senior engineering director Neha Batra describes mentoring as (*https://oreil.ly/2QWel*) "sharing your experience so an engineer can leverage it themselves." But remember, mentoring is focused on *your* experience.

Solicited or not, the advice that worked for you might not work for someone else. Author and management coach Lara Hogan (whose name will come up a lot

---

2 Don't name or shame the person you think needs advice, though.

this chapter!) warns that "advice that might work for one person ("Be louder in meetings!" or "Ask your boss for a raise!") may undermine someone else, because members of underrepresented groups are unconsciously assessed and treated differently."[3] The best practices of a decade ago also might not work for a younger coworker now, and the social dynamics (or technology stack) of the experience you had might not map well to the one your colleague is facing. I've had colleagues push back on advice that I targeted wrong. "Just DM the director and ask them to invite you to the meeting" is an easy thing to say when the director is a peer, and much more difficult when they're your boss's boss's boss!

Be careful of unsolicited advice in mentor/mentee conversations. When someone starts describing a difficult project, for example, the easy, intuitive response is to say what you, the sage advice-giver, would do in the same situation. It's kinder (if more difficult) to figure out what they actually need. Maybe it turns out that they do want help. But it's just as likely that they're looking for reassurance that other people would also find the problem or situation difficult. They might be doing rubber duck debugging (*https://oreil.ly/vCJot*), explaining something to you so they can unpack it for themselves. They might just want to tell a war story, to hear commiserations and congratulations for what they've navigated so far. Unsolicited advice derails all of those things. It can help to ask "Do you need space to vent or are you looking for advice?" and then give either comfort and validation or solutions as requested.

Mentoring is not just for new people: I have mentees with decades of experience as well as mentors of my own who I ask for advice. It's not necessarily one-way either. Your mentee might give you a new perspective or teach you about topics that they know better than you do.

If you're getting into a mentoring relationship, set it up for success. Set expectations, such as that you'll meet once a week for six weeks. Agree on what you're trying to achieve: does the mentee want to onboard and feel comfortable in a new company, learn a new codebase, or get career advice to strive toward a new role? If all of this is settled up front, you're less likely to find yourselves sitting in a room staring at each other all "What were we supposed to talk about?"

---

3 *Resilient Management* (A Book Apart)

Answering questions

If you have a vast repository of knowledge and everyone's afraid to ask you anything, your knowledge will stay in your own head. Be accessible. Depending on
your work style, that might mean offering office hours, being friendly and easy to
DM, or spending some of your time just hanging out near the teams you work
with—office spaces with sofas are fantastic for this sort of thing.

Make reaching out to you worth the effort. Some engineers seem to guard
their knowledge preciously, answering only direct questions and not a
word more.

> *"Will the /user endpoint give me the user's full name?" asks the junior
> engineer.*
>
> *"No," replies the senior engineer, "it can only give you the username."*
>
> *The junior engineer tries to hack around the problem for an hour before
> nervously asking, "Is there a different endpoint that could give me the full
> name?"*
>
> *"Sure," says the senior engineer, "use /fulluser."*

What a waste of an hour! The senior person had information, and it didn't
occur to them to impart it. That doesn't mean you should *infodump*, offloading
every snippet of information that could connect to the topic at hand. But understand what advice is being *implicitly* solicited, even if it's not directly asked for. If
you're not sure, ask what your colleague is trying to do and ask if they need help.

Code and design review can be another time to answer implicit questions
and give advice. If you see a place where your colleague could solve a problem in
a better way, tell them. But be clear about whether you're just sharing interesting
information or asking them to change course: it's frustrating to get a bare comment like "The Foo library also would work here" without context about whether
that means you hate the current approach.

Domain-specific information goes beyond technology. If you've learned ways
to navigate your organization and get things done, that's valuable knowledge that
you can pass on to mentees and other colleagues. Share the topographical map
you built in Chapter 2!

Feedback

One of my self-appointed roles in my current job is to be a test audience for colleagues doing conference talks. I love watching tech talks and I've invested a lot

of time in learning how to do them well, so the presenter and I both benefit. As I watch the talk, I take a ton of notes, highlighting the parts I found funny, insightful, or educational. But I also point out anything that didn't work, anything I thought wasn't correct, anywhere I started to tune out. Good and bad, I tell the truth. It's a waste of the presenter's time otherwise.

When someone asks you to review a document or pull request or conference talk, do call out the sections that you think are great, but pay them the respect of being (kindly!) honest about their work. Giving constructive and critical feedback isn't easy. It takes effort to tease out exactly what isn't working and find the words to explain why it's not as good as it can be. It's a more difficult conversation. But you won't help your colleague if you hide the truth.

Peer reviews

If your company has a performance management cycle, you might be asked for a specific kind of feedback: peer reviews. These reviews have two audiences and you should keep both in mind.

The first is the person who asked you for feedback. Assume that they asked because they genuinely want to know how to improve. I've seen people struggle with "What could this person do better?" questions in peer feedback, because it's easy to see the answers as criticism. But take the question literally: what *could* your colleague do better? How could they become more awesome? If you can't think of anything, ask yourself why they aren't one level more senior (or two!), and give them advice on behaviors they should focus on to get there.

But also remember that the feedback will be read by the person's manager and potentially others who are calibrating their performance or evaluating them for promotion. Give those people the information they need to help your colleague grow and to notice patterns that need to be addressed. But think about how your words will be perceived and whether they can be taken out of context by someone who doesn't know the work you're describing. If you find yourself holding back on describing an area for growth because you don't want to accidentally torpedo a well-deserved promotion, consider delivering private feedback by email or in person instead.

Just like giving mentorship advice, remember that what works for one person might be terrible advice for another. This is especially prevalent in advice about communication style. For example, "be aggressive" is advice that will make some people seem more like leaders, but will get others in trouble. It's a common joke in tech women circles that you know you're acting at senior level when you get your first peer review saying you're "abrasive."[4] (That might also be the first time your reviews don't say you should be "more assertive." It's a fine line to walk.) So watch out for implicit bias (*https://oreil.ly/KYbCm*) and be aware of how you're describing the same behavior across different people. Was it "consensus building" or "indecisiveness"? Were they "refreshingly down to earth or "unprofessional"? Often it depends on who you're talking about. The folks at Project Include offer more recommendations for providing feedback (*https://oreil.ly/UMO42*).

SCALING YOUR ADVICE TO A GROUP

You can't meet individually with everyone who needs advice or write feedback for your whole organization: you'd have no time left in your quarter. But you can give a tech talk or write an article on literally anything you want, and chances are it will reach some people who find it helpful. That's a great way to scale your advice.

Years ago, a group of volunteers at Google wanted to increase the quality of testing and hit on a novel solution: they started writing their advice as simple one-pagers, printing them out, and putting them up in toilet stalls. "Testing on the Toilet" (*https://oreil.ly/F2bnz*) is the ultimate in unsolicited advice, but it's popular and amusing, and people read it!

If you want to tell something to more people, write it down. Documentation means that you don't have to explain how to do something again and again. A FAQ, a how-to, even a descriptive channel topic can let you say something once but have it read by many people. And if you write something that applies to people outside your company, consider sharing it further as a blog post or article.

In addition, look for opportunities to get a microphone and an audience. Can you get a slot at an all-hands meeting, conference, or tech talk event? Sometimes you can use these opportunities to deliver a message you're focused on *alongside* the message you've been asked to share. I was invited once to present to a huge group about an incident of my choice. At the time, I was really trying to get the word out about dependency management and how important it is to be deliberate about what systems you depend on. So I talked the group through a recent outage that had been made much worse by a circular dependency that

---

4 2014. A good year.

stopped the systems from coming back online.[5] The opportunity to present at this big meeting let me frame the outage story so that it highlighted the message I wanted to share. It was an audience and a microphone at the right time.

BEING A CATALYST

If you want to be a catalyst, set up advice flows that don't need you to be involved. Make it easy for your colleagues to help each other.

If your team or organization relies on one-on-one conversations to understand how anything works, one of the most powerful things you can do is encourage people to write things down. Start small: going from an oral culture to writing *everything* down won't win you any friends. Instead, look for a small but meaningful change. Are you missing an easy-to-use documentation platform? Might teams sign on for creating a FAQ of the most common questions they're interrupted by? Might your director endorse a quarterly documentation day?

You can scale audience-and-microphone-style advice by setting up monthly tech talks or lunch-and-learn meetings. This is a bigger commitment than just scheduling the meeting: you'll have to solicit talks, send reminders, and possibly watch practice runs. Be clear about what you're getting into, plan for it on your time graph, and ideally start with at least three people to share the load.

Similarly, you can scale mentorship further by setting up a mentorship program. Be warned: the administrative work will be more time-consuming than you might expect, and this work is generally not considered to be part of an engineer's job. If you can find a manager who is interested in doing the same thing, they will likely find it easier to frame the work as part of their job description. Convincing *someone else* to set up a mentorship program still counts as being a catalyst!

Teaching

On to the second type of good influence, and it's a step up from advice: teaching. What's the difference between *telling* people things and *teaching* them things? Understanding. When you're giving advice, you're explaining how *you* relate to the topic, and the receiver can take your advice or leave it. When you're teaching, you're trying to have the other person not just *receive* the information but internalize it for themselves.

---

5 X couldn't start without Y and Y couldn't start without X, and so neither could start.

Deliberate teaching is not just for a more senior person to help a more junior one: it's useful any time someone new is joining a team, or when you've got more domain knowledge than someone else. I love asking a colleague for an overview of their systems, for example, so I can fill in a mental gap in the overall architecture. At the end of an hour of whiteboarding, I'll have a much clearer picture of how their systems interact with others, and I'll have made my own diagram to refer back to or to add to their documentation.

INDIVIDUAL TEACHING

Anywhere there's a knowledge gap between you and someone you're working with, there's an opportunity to teach. Sometimes this means formal training or coaching. But there's also plenty of teaching to be done within the regular structures of working together: pair programming, shadowing, and review.

Unlocking a topic

Think back to the best classes you ever took. What was successful about them? I bet you walked away feeling like you had a handle on something you didn't have before: a new skill or understanding that you could build on. Great classes "unlock" a topic for you, sparking curiosity and interest.

Teachers have a specific goal, often formalized in a lesson plan. If you're teaching, you should too. Some examples:

- Are you giving an overview of a system? If so, by the end of the session, the person you're teaching should be able to draw the system on a whiteboard and describe it to someone else.

- Are you walking through a codebase? Aim to give them everything they need to send their first pull request.

- Are you showing them how to use a tool or API? Describe three to five common scenarios they should be able to handle on their own by the end of the session.

Successful teaching includes hands-on learning and activating knowledge: the student should be *doing* as well as listening. Find opportunities to let them lead, whether that means using the tool themselves, typing commands, or opening tabs on their laptop instead of yours. It's even better if they end up with an "artifact" to refer back to, like a diagram or code snippet.

Pairing, shadowing, and reverse shadowing

Here's another way you can teach: working directly with someone else. Working together has a spectrum of approaches (see Figure 8-2), from shadowing, where you're doing all of the work with your coworker observing, to pairing, where you're working together, to reverse shadowing, where *they're* doing all of the work with you watching to give them feedback.

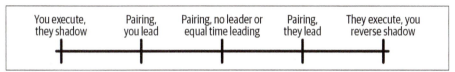

Figure 8-2. The spectrum of working together. Different points on this line will be useful in different situations.

Shadowing is a way of teaching by demonstrating: your "shadow" watches you execute a skill and takes notes on how you're approaching it. It's a great opportunity to be a visible role model and show your colleague how to work with high standards.

With pairing, you're still working together, but the "shadow" has become an active participant. Pairing can mean pair programming, coauthoring a document, whiteboarding an architecture, or solving a problem together. It's another opportunity for role modeling, but also for teaching: working side by side means you'll be able to share knowledge and check for understanding in real time.

Finally, you might "reverse shadow," where the learner performs the task and the experienced person watches and gives notes. No matter how closely the learner has paid attention, they'll learn the most by activating the knowledge and practicing the task. Reverse shadowing can also serve as a type of guardrail, which I'll discuss later in this chapter.

Code and design review

Reviewing code and designs can be an excellent form of teaching. You get to highlight perils your colleague might not know about and suggest safer alternatives. You also get to encourage behaviors you want to see more of.

A review from a senior person can be a real confidence booster. Be careful, though: a review done *wrong* can destroy someone's confidence rather than boosting it. It can be soul-destroying to work through a barrage of comments that are condescending, seem arbitrary, or that you don't understand.

As a teacher, your job is to impart the knowledge and point out problems in a way that retains your student's confidence and growth mindset. Code review will show up again in the "guardrails" section, and I'll talk there about reviewing to prevent harm to your systems, but for now, here are some ideas to bear in mind as you review to *teach*:

Understand the assignment   Be aware of the context. Is your colleague new to the language or technology and looking to learn, or do they just need a second pair of eyes for safety?[6] Understand the stage of work, too: if you're reading a first high-level draft, start with the foundations and the approach and don't get into the nitpicky details. If everyone has bought in and this is the last review before launch, it's not the time for big directional questions: get right into the weeds and be extra alert for what could go wrong.

Explain why as well as what   A review comment like "Don't use *shared_ptr,* use *unique_ptr*" only tells the code author what to do right now. They won't know what to do next time. Teaching means sharing understanding, not just facts. While the code author can go read documentation on whatever you just told them about, they might not recognize why it applies. A short explanation or a link to a relevant article or specific Stack Overflow post (rather than a general manual) will be a shortcut to help them learn.

Give an example of what would be better   If a section of a design is confusing, don't just say "please make this more clear." It's hard to know what to do with that! Offer a couple of suggestions of what you think the author is trying to say.

Be clear about what matters   When you're less experienced, it can be hard to calibrate the advice you're given. Some things are vitally important, some are nice to have, and some are just personal preference. Annotate your advice so it's clear. Some examples:

---

6 I've sometimes tried to anchor reviewers by noting the context in the change description. For example: "I'm new to this language, so if something looks weird, it's probably not a deliberate stylistic choice! I welcome nitpicking and advice on what's idiomatic."

- "Use parameterized queries here instead. You're opening yourself up to SQL injection attacks—a malicious user could drop our database!"

- "The way you're approaching this will work fine, but we prefer to avoid the singleton pattern. Here's a link to the section of our style guide that talks about why."

- "I'd recommend one bigger microservice here rather than two small ones —I think it'll be easier to maintain. But I don't feel strongly; your call."

- "This is just a nitpick, but all of the other spellings in this file are in American English, so let's call this one *organization* instead of *organisation*.[7]

Choose your battles John Turner, a software engineer who has written about code review for the Squarespace engineering blog, recommends reviewing code in several passes (*https://oreil.ly/3VLqa*): first high-level comments, then in increasing detail. As he points out: "If the code doesn't do what it's supposed to, then it doesn't matter if the indentation is correct or not." This advice works for RFCs too: if your first comment is that the author is solving the wrong problem, it's not helpful to leave a hundred technical suggestions.

If you mean "yes," say "yes" Make it clear whether you consider your comments to be blocking or not, and whether you're otherwise happy with the change. Call out the good as well as the bad. In particular, explicitly say "This looks good to me" on design documents. Code review tends to end by clicking a button to say that you believe the change is safe to merge. When there are a lot of reviewers, though, each one may be hesitant to approve until the others have weighed in. If you have no objections, say so.

Remember that engineers who are earlier in their careers may find you intimidating or be reluctant to question your suggestions even when they think you're wrong. Think about how to make your comments friendly, approachable, and, well, *human*. If the pull request or RFC needs a lot of work, the most constructive approach might not be to bury your colleague in comments: consider setting up time to talk or pair with them instead.

---

7 A little window into my life. :laughing:

Coaching

The last form of individual teaching I'm going to talk about is coaching. Whereas *mentoring* involves sharing your personal experiences, *coaching* teaches people to solve problems for themselves. It can be a slower process, but your colleague will learn much more by making their own connections than they will from you giving them the answers.

Coaching's a set of odd skills that sound pretty straightforward but take time and deliberate effort to learn. You shouldn't expect to be immediately good at it! Here are the three big skills you'll need:

Asking open questions Open questions are the ones that can't be answered with "yes" or "no": they yield much more information. Try to dig into the problem rather than starting with a solution. Ask questions that help your coachee identify and unpack aspects of the problem that they might not already have considered.

Active listening Reflect back what you've heard, to make sure that you've really understood and to let the person hear how you frame what they've said. The feeling of being understood can be powerful and help people feel less alone. Your framing can give your colleague a new way to describe what they're working through, helping them come up with new solutions.

Making space Leave enough space and silence for the coachee to reflect. If you tend to reflexively jump in when there's a silence, count to five in your head before speaking again so the other person can process.

When you're new to being a coach, it feels very strange to not just *help*. If you have the answer, shouldn't you just *give* it? No. As management consultant Julia Milner warns in her TEDx talk (*https://oreil.ly/ghwkC*), you can't know every detail of the situation. When you provide a solution, the coachee is likely to reflexively respond with what Milner calls a "Yes, but...," a reason your advice can't work. Instead, she says, good coaching involves drawing out their own best ideas, providing them the space to reflect, and helping them take their own journey to a solution that works for them.

SCALING YOUR TEACHING TO A GROUP

Teaching one person at a time is great for that one person, but it's a slow way to spread information. You can scale your teaching by creating materials for a class.

Putting together a class takes a ton of effort. There's a high up-front cost for the first time you teach it, but you'll amortize that cost every time you teach it again. Just like individual teaching, your class should have a specific goal for what your students will know or be able to do at the end. Include exercises or some way for the students in the class to practice what they've learned.

If you want to make your class asynchronous, consider making something that students can use at their own pace. A great example of this kind of teaching is a *codelab*: a guided tutorial that takes the student step by step through building something or solving exercises.[8]

I used to work on a project that was perpetually underfunded and survived on a rotating cast of volunteers, interns, and people on short-term residency programs. Many were new to the company or even the industry. Dropping them straight into our codebase (an intimidating networking library) would have made them run screaming, so we created and documented a learning path.

On day one, we had them send two pull requests: one to add a joke to our repository of jokes, and one to add their name to the list of people on the team. We would find a reason to do a little back-and-forth on this, so they'd get used to the code review process. On day two, we'd have them build and run a tiny client and server we'd created just for this purpose. They'd watch it running in production, and look at its UI and the logs and metrics it was exporting. Then they'd make a local change to the library—adding a new log message, maybe, or even tweaking the logic—and deploy it to prove to themselves that the code still worked and that they could see their change in the monitoring data. It was very effective. By the time they got to making actual changes on week two, they weren't scared of the code. It was just code.

When the majority of your contributors only stay for three months, you need to get them up to speed quickly. We used this same well-documented learning path for every new contributor, and it didn't take long for it to repay our investment.

BEING A CATALYST

You can scale your classes further by teaching other people to teach them. Different teachers have different styles; embrace that. Let your new teachers begin to own their own classes: they should have access to edit the slides and exercises, or they should have your blessing to create their own variants. Shadow them and give honest feedback: they want to learn. Once *they* start teaching other people to

---

8 The Kotlin Koans (*https://oreil.ly/XXtus*) are a great example, and fun to work through. Google also has a ton of great codelabs (*https://oreil.ly/OqlOA*).

teach the class, it has life without you. (At this point I usually slink away and remove myself from the teaching rotation.)

If your class is applicable to all your engineers, try to add it to an onboarding curriculum or internal learning and development path—all of your new hires will learn what you want them to know without you having to find a way to reach them. If your company doesn't already have this kind of learning culture, you can have a huge impact by advocating for an onboarding process, evangelizing learning paths, or setting up a framework that makes it easy to create codelabs.

Guardrails

Think of the railings you might find along a cliffside walking path. They're not for *leaning* on, but they're there to steady yourself when you need them. A small stumble won't doom you: the railings will stop you going over the edge. Guardrails encourage autonomy, exploration, and innovation. We all move faster when the going is safe. In this section we'll look at some ways you can add guardrails for your colleagues, first individually and then at scale.

INDIVIDUAL GUARDRAILS

You can provide guardrails by reviewing code, designs and changes, and by offering support through scary projects.

Code, design, and change review

I've already described how code and design review can be great teaching tools. There's a third type of review that can be effective, too: *change management*. That's the process of writing down exactly what you're going to do before you do it and having someone else agree that you've described the right set of steps. It's like code review, but for command lines or clicking buttons in the right order.

You can use code, design, and change reviews as powerful guardrails to help your colleagues. When someone knows their work will be reviewed, it's easier for them to feel confident working independently. They know there will be a check to make sure they don't cause an outage or spend months building an architecture that can't work. The guardrail helps them avoid dangerous mistakes.

If you want to be a good guardrail, don't ever rubber-stamp changes. Read carefully: every line of code, every section of a design, every step of a proposed change. Here are some categories of problems you should look for:

Should this work exist?

What problem does your colleague intend to solve? Are they using a technical solution to solve a problem that should have been solved by talking to someone?

Does this work actually solve the problem?

Will the solution work? Will users be able to do what they need and what they expect? Are there errors or typos? Any bugs or performance issues? Does the design propose using a system in a way that won't work?

How will it handle failure?

How will the solution handle weird edge cases, malformed input, the network randomly disappearing, load spikes, or whatever else can go wrong? Will it fail in a clean way, or will it corrupt data or take a user's money without giving them the service they've paid for? How will you discover problems?

Is it understandable?

Will other people be able to maintain and debug new code or systems? Are the components or variables named intuitively? Is the complexity contained in a well-chosen place?

Does it fit into the bigger picture?

Does the change set a precedent or create a pattern you might not want other people to copy? Does it force other teams to do extra work for future changes? Is this a risky change that's scheduled at the same time as a high-profile launch?

Do the right people know about it?

Is everyone copied on the change who should be? Are there names attached to any actions that need to happen, or is there a lot of passive voice where it isn't clear which team is doing what? Do the people involved know what is expected of them?

As a reviewer, be open to the idea that you don't know everything. Ask questions and be constructive. A good guardrail is not an arbitrary gatekeeper: you're on the same side as the person you're keeping safe, and you want them to succeed.

Project guardrails

If you've ever stretched to take on a difficult project, you'll know that it's a great way to build skills, but it's also nerve-wracking! You're more likely to fail, because you're doing something that's hard for you. So it's nice when you have a more experienced colleague who has done a project of this size or shape before: they can help keep you safe. That doesn't mean they'll do the work for you or protect you from all possible mistakes, but they'll let you know if you're getting close to a disaster you won't be able to recover from. That's what it means to act as a project guardrail.

I remember leading a project that was a real stretch for me. It had more moving parts than anything I'd done before, more stakeholders, and *way* more politics. Every week when I met with my team lead, he'd ask questions about the project: "Just out of interest, how were you planning to balance these two conflicting business priorities?" Of course, I didn't have a plan: I hadn't noticed the problem creeping up. But the questions were enough to put me back on course, and I was able to suggest paths forward. Although I didn't realize it at the time, the team lead was acting as a guardrail. He was making sure I'd noticed that I was walking close to a cliff edge, and if I didn't have a good idea for what to do, he was ready to coach.

Being a project guardrail isn't just for less experienced folks: you can play this role for any colleague who is leading a project or taking on a difficult task. Even very seasoned people can use support in a project that's using a new skill set. If someone asks you to be a mentor or adviser on a project, they're probably hoping for at least a little guarding.

A guardrail can offer support as well as safety. Lara Hogan suggests being specific (*https://oreil.ly/gACoX*) about how you can help with the project, like promising to review designs or advocate for ideas with upper management, as well as being explicit about when and how your colleague should ask you for help. She suggests lines like "Shoot me an email if person B is unresponsive to you for three days; I can be your muscle there."

SCALING YOUR GUARDRAILS TO A GROUP

You can't personally review *every* change and support every project, and you'll just slow everyone down if you try. Let's look at how you can add guardrails for your team or organization without getting in everyone's way.

Processes

Rather than individually teaching your coworkers the right thing to do, you can write down a standard set of steps and convince the organization to follow them. For example, what's the "right way" to launch a new feature at your company? Your prelaunch process might include answering questions like:

- Do we need security approval?
- How much notice should we give the marketing team or customer support?
- Should we ship behind a feature flag?
- Is there standard monitoring, eventing, and documentation we should add?
- Do we need to tell other teams to expect extra load?

And many more! As the company grows and the organization gets more complex, there will be more ways for a launch to cause an outage or public relations mess. And so a process is born.

Opinions about processes vary. Some people will be delighted to have clear steps and the safety that comes from standardization. Others will insist that people should *think* instead of mindlessly following protocols and that checklists and approvals just slow them down. Nobody's wrong; it's a trade-off. But the bigger the company, the more likely you'll need *some* sort of structure that helps people do the right thing without having to ask the same questions every time.

Here are some other examples where adding a process or checklist might be helpful:

- Responding to a major outage or security incident
- Sharing and agreeing on RFCs or designs
- Adopting a new technology or language
- Making and recording decisions that cross multiple organizations

In general, aim to make processes as lightweight as possible. If you add a complicated procedure with lots of boilerplate, central approval, and long waiting periods, it's not going to be a good guardrail: people will just sneak around it. Make the right way the easy way.

Process Preamble

Here's the introduction I wrote for a process FAQ document at work. Feel free to use it if it's helpful for you too.

There are a lot of questions about how <topic> should work. It's hard to find a balance for how prescriptive to be with processes like this.

- If you write nothing down, most people hate that and complain that they don't know how to do anything.

- If you write down guidelines, people interpret them as law and argue that they're wrong because they don't cover edge cases.

- And if you write down every edge case, you end up with a three-ring binder of policy and legalese, and it probably still won't cover every situation. And everyone still hates it!

This document attempts to give mostly correct answers to some frequently asked questions. These answers will not apply perfectly in every situation. Think twice before discarding them, but if they don't make sense for a situation you're in, do the thing that makes sense instead. All guidelines are wrong sometimes. (If these guidelines are wrong *a lot,* propose a change.)

When in doubt, think hard about the other humans involved in what you're doing, assume they're reasonable people trying their best, and also be a reasonable person trying your best.[9]

Written decisions

Here's another way you can make it easier to do the right thing: make a decision once and write it down, so people don't have to have the same argument again and again. Written decisions remove a little decision fatigue from people's lives: the rules say we usually do X, so that's what we'll do!

Here are four examples:

Style guides As Google's style guide site explains (*https://oreil.ly/gkUmT*), a style guide for a project is "a set of conventions (sometimes arbitrary) about how to

---

9 This is also just generally good life advice.

write code for that project. It is much easier to understand a large codebase when all the code in it is in a consistent style." The word *style* here covers a lot of ground, from naming conventions to error handling to which language features are OK to use. By making the decisions once and writing them down, you save teams from having the same "do we use lower camel case or snake case for variable names?" arguments for every new project. You'll end up with more consistent code, too.

Paved roads Some companies document their set of standard, well-supported technologies and recommend (or mandate) that teams don't step off that "paved road." I like the format popularized by the Thoughtworks Tech Radar (*https:// oreil.ly/FUHQP*), marking technologies as "Adopt," "Trial," "Access," and "Hold."

Policies Companies can make rules: for example, "Every team should run a retrospective after an outage." If the rule is enforced, breaking it could be seen as failing to do one's job, with implications for performance reviews. Use policies sparingly. It's hard to account for all the edge cases—and there will always be edge cases. Besides, if there are too many policies, people just won't remember all the things they're supposed to do.

Technical vision and strategy A technical vision or strategy (see Chapter 3) gives a clear direction within which teams can choose their own paths to solving problems.

Robots and reminders

Software consultant Glen Mailer says he looks for ways to make it as easy as possible for people to remember to do the right thing. This means putting the right solution in their faces—sometimes literally! He gave an example of a workplace where everyone was supposed to track their project time using timesheets. Of course, people often forgot until someone came up with a solution: they stuck a timesheet grid and a pen on the exit door at head height. When anyone pushed the door to leave, the timesheet would be in front of their eyes—much harder to forget.

If you're trying to introduce a process or a written decision, see if there are ways you can (gently) put it in people's faces. Even better, have an automated system do the right thing so humans don't have to. Some examples:

Automated reminders Rather than always reminding someone that it's their week to follow the release process, set up automation that puts it in their calendar or DMs them about it. The reminder should include a link to the process.

Linters Have a code linter enforce as much of your style guide as it can, so reviewers don't have to.

Search Make sure that any search for how to do something brings up the *right* way to do it, even if that means updating all of the "wrong" documents to have headers that point to the right place.

Templates If all RFCs are supposed to have a security section, make sure there's an easy RFC template and that it includes a security section.

Config checkers and presubmits Can you add automation that automatically runs unit tests, or runs safety checks on configs before committing them? Google's data center safety system, SRSly, is a great example: it allows setting guardrails like "No more than 5% of servers may be rebooted at once" and "Don't decommission a server for this system if the on-call for it recently got paged."[10]

BEING A CATALYST

Creating robots, policies, and processes that reinforce your message scales further than being a guardrail for individual colleagues. But they all still rely on *you* doing something. If you really want a message to stick, you need everyone to believe in it and care about it. You want your organization to get to a state where it would be considered weird to do something else. The most effective guardrail is also the most difficult to put in place: culture change. Unless you can make the guardrail part of your culture, you'll always be chasing compliance.

Most tech companies now have code review and write tests. But that wasn't always the case! All of the guardrails that we take for granted today were introduced by people who cared enough to argue for why the change was worth the time. If you're introducing a culture change, be patient. It takes a lot of time and dedicated effort to make everyone behave in a different way, but it's the only way you'll ever be able to stop pushing the process along manually.

Here are some ways you can make your culture change journey easier:

---

10 Check out Christina Schulman and Etienne Perot's entertaining talk (*https://oreil.ly/k8ohr*) on SRSly, including its origin story: automation accidentally sent all of the disks in a data center to be erased at once. As they noted, if you ask efficient automation to do something *stupid*, it will do so very efficiently.

Solve a real problem

The culture change should be closely aligned with what the organization needs. Expect any proposal to be confronted with a lot of "why" questions. Have good answers that aren't just aspirational: really, what does the business get out of this?

Choose your battles

Rather than a process for design review, try to instill a *respect* for design review and trust that teams will make their own choices about what form works for them. Offer some easy defaults, but don't get hung up on whether everyone's following exactly the same process.

Offer support

Your processes and automations should support the change you want: they should make it easy to do the right thing.

Find allies

Don't try to change the culture on your own. Ideally, your allies will include high-level sponsors and influencers in the organization you're trying to change. Consult the *shadow org chart* you mapped out in Chapter 2.

Opportunity

The last type of good influence we're going to look at is finding people the experiences they need to grow. People learn by *doing*, even more than they do from teaching or coaching or advice. Every project or role is a chance for visibility, relationships, and résumé lines, all of which can lead to further opportunities. Let's look at how you can send those experiences to your colleagues, both individually and at scale.

INDIVIDUAL OPPORTUNITIES

As a senior person, you will have many occasions where you can help other people find opportunities to grow. You'll be able to directly offer projects and learning experiences through delegation. But you'll also be able to suggest people for assignments, promote their work, or connect them with information that can help them.

Delegation

Delegation means giving part of your work to someone else. When you delegate, you're usually not just tossing someone a project and walking away: you're invested in the outcome. That might mean you're tempted to micromanage or to handle all of the difficult parts of the project yourself. But when you hand over the work, *really* hand it over. As Lara Hogan says (*https://oreil.ly/PJo6s*), your colleagues won't learn as much if you only delegate the work after you've turned it into "beautifully packaged, cleanly wrapped gifts." If you instead give them "a messy, unscoped project with a bit of a safety net," they'll get a chance to hone their problem-solving abilities, build their own support system, and stretch their skill set. A messy project is a learning opportunity that's hard to get otherwise.

Target the level of difficulty to the person you're delegating to—don't throw organizational chaos to a new grad! But when you're looking for someone to delegate to, think beyond the most obvious people. Anyone who can do an A+ perfect job on a project isn't going to learn from it.[11] Instead, try to find someone who will find the work a bit of a stretch but manageable with support. Promise them that support. You might need to give them a little push to help them see that the work is within their reach: they might not yet see themselves as a project lead, an incident commander, etc., but the fact that *you* see them like that can be a tremendous boost to their confidence. Describe the guardrails you can provide for them and explain why you think they can handle the project.

Be warned that, when you delegate, you're *not going to get a clone*. (Sorry, we don't have that technology yet.) Inevitably the person you've delegated to is going to take a different approach than you would have. Be a guardrail, coach them, and ask questions, but inhibit the urge to step in. So long as they're going to achieve the goals, let them do it their way. I absolutely love how Molly Graham, most recently the chief operating officer at Quip, frames handing off work as "giving away your Legos" (*https://oreil.ly/ltqKy*):

> There's a lot of natural anxiety and insecurity that the new person won't build your Lego tower in the right way, or that they'll get to take all the fun or important Legos, or that if they take over the part of the Lego tower you

---

11 This pattern is common in recruiting mails: "Come do exactly the thing you're currently doing, but at another company." There are times when that will work (we'll explore motivations for changing jobs in Chapter 9), but the most successful recruitment I've seen is for roles that offer people a step up, something slightly scary.

were building, then there won't be any Legos left for you. But at a scaling company, giving away responsibility—giving away the part of the Lego tower you started building—is the only way to move on to building bigger and better things.

One key Lego-relinquishing behavior to watch out for is that you should be *redirecting* questions about the project to the other person, not *proxying* the information. That is: when someone asks you a question about the project, you may think that you have the current state. But answering the question makes you the point of contact, potentially undermining the colleague to whom you handed off the project. You might also have the wrong answer! Instead, give visibility to the person who took over the project: note that they're the expert and owner for the topic and show that you trust them to make decisions. You'll give the project owner a connection to someone new and any extra opportunities that come out of that connection. The next time the interested person has a question about the project, they'll know where to go. And the project will be off your plate.

Sponsorship

Sponsorship is using your position of influence to advocate for someone else. It's more active than mentorship: you're deliberately unlocking opportunity for other people, not just giving them advice. It takes more work, too: if you want to be a great sponsor, you need to know what your colleague will benefit from and what opportunities they're looking out for. You're investing your time and social capital in their growth.

Rosalind Chow, associate professor of organizational behavior and theory at Carnegie Mellon University, has described what she calls "the ABCDs of sponsorship" (*https://oreil.ly/Ndm87*):

Amplifying
> Promoting your colleague's good work and making sure other people know about their accomplishments

Boosting
> Recommending them for opportunities and endorsing their skills

Connecting
> Bringing them into a network, giving them access to people they wouldn't otherwise be able to meet

Defending
> Standing up for them when they're unfairly criticized; changing any negative perceptions of them

The opportunities that you can offer through sponsorship may seem like small ones, but they lead to greater things. If you recommend someone to lead a small project, you're setting them up to later be seen and chosen for a bigger one. Every time you give someone a shout-out, comment on their work, or even retweet them, you're signal-boosting their good work and making sure the other people in your network know about them and think of them when opportunities arise. If someone in another team is being a superstar, make sure their boss knows. If you're in a company that writes peer reviews, a review is a great way to make sure the good work goes on the record.

Who should you sponsor? Look for people who want opportunities to grow and who you trust to do good work. You're spending your social capital by recommending them, so don't waste that by sending opportunities to people who don't actually want them, or who won't put in the effort. Sponsor colleagues who have untapped potential, who are worth investing in. Watch out for in-group favoritism, though, a cognitive bias that can let you evaluate other people more favorably if they're like you. In the words of Mitch Kapor (*https://oreil.ly/bbO1B*), founder of Lotus, cofounder of the Electronic Frontier Foundation, and cochair of the Kapor Center for Social Impact: "We talk about the meritocracy of Silicon Valley, when really it's a mirrortocracy, as people tend to hire people who look like themselves at greater rates than other sectors." Pay attention to who you're recommending or helping, and make sure you're not accidentally only sponsoring people who look like you. It's surprisingly easy to do.

Connecting people

Even if a role or project isn't yours to delegate and you aren't being asked to make a recommendation, you can still offer opportunities just by knowing that they exist. As a staff+ engineer, you'll probably have broader context than other engineers you work with: as I described in Chapter 2, you'll be spending more time just *knowing things*. Keep an ear out for opportunities that will help your colleagues. You can remember that a conference call for papers is open, that a team lead role is opening up, or that a new internal training program is offering a skill someone is trying to learn. By connecting people with information, you expand their options.

SCALING YOUR OPPORTUNITIES TO A GROUP

Some of the catalytic types of influence I've already described give people visibility and opportunities to learn leadership and teaching skills. But here's one more way you can scale opportunity: sharing the stage.

Share the spotlight

As the most senior person on a team or on a project, there's going to be a lot of things that you do best. That might make it tempting to jump on every difficult problem. But while it might feel amazing to be the resourceful and knowledgeable senior person who's leading the team to greatness every day, what you're actually doing is overshadowing the rest of the team and preventing them from growing.

Instead, translate your superstardom into helping *everyone* do better work. Let other people do work that's not as good as you would have done it, so long as it's *good enough*. That's how they learn. Sharing the stage includes delegation, but it can also mean making space: letting other people notice that the work needs to be done so they can build leadership skills by picking it up or learning to delegate for themselves.

Here are some ways you can make sure you're sharing the spotlight:

- If someone asks a question in a group meeting, leave a gap or explicitly hand off the floor to another person on your team.

- Add a less senior colleague to a meeting you go to, and let them speak about their work.

- Invite a less senior colleague to review designs or code that connects to their work, and be clear that their opinion matters.

Warning

While being the face of every project limits your team, the opposite extreme can have problems too. If you prefer to delegate *everything*, make sure you're aligned with your management chain about how you work. Most managers will expect some direct execution and visible accomplishments from their staff engineers. Don't put yourself into so much of a support role that nobody's quite sure what you do.

BEING A CATALYST

You can take steps to keep opportunity and sponsorship flowing even when you're not there. While most workplaces have some concept of mentorship, sponsorship might not already be well understood. Look for ways that you can teach your colleagues to look beyond the obvious candidates to suggest for opportunities. Some ideas include: advocating for an inclusive interview process (*https://oreil.ly/PdxHq*), inviting a speaker on implicit bias, or setting the culture that open team roles are posted on an internal job board.

The best way you can be a catalyst in the industry is by empowering your colleagues to become great engineers who do great things. As you offer opportunities that turn your midlevels into seniors and then into staff engineers, teach *them* to offer advice, teaching, guardrails, and opportunities to the people who will follow them. The staff engineers you grew will grow the staff engineers who follow them. Your leadership will keep going.

Warning

For someone to be promoted to your level, they don't need to be as good as you are *now*: they need to be as good as you were when you were first promoted to the level. If you keep seeing people join your level who aren't as capable as you are, don't snark about lowered standards: think about whether that means you've grown.

At the start of the chapter I listed three reasons to improve your coworkers' skills: getting more done, keeping your technology up to date, and improving the industry. Here's a fourth: other people's growth is your growth. If you can delegate, you'll be able to take responsibility for bigger, more difficult problems, handing off parts of them to the rest of your group. The more your colleagues can do, the more you can do. As Bryan Liles says (*https://oreil.ly/ScVHa*), "How you can get pushed up is by building a whole bench behind you."

At some point, you might look at the people who are doing the job you used to do and realize you've gone up a level. What do you do then? In Chapter 9 we'll look at what's next.

To Recap

- You can help your colleagues by providing advice, teaching, guardrails, or opportunities. Understand what's most helpful for the specific situation.

- Think about whether you want to help one-on-one, level up your team, or influence further.

- Offer your experience and advice, but make sure it's welcome. Writing and public speaking can send your message further.

- Teach through pairing, shadowing, review, and coaching. Teaching classes or writing codelabs can scale your teaching time.

- Guardrails can let people work autonomously. Offer review, or be a project guardrail for your colleagues. Codify guardrails using processes, automation, and culture change.

- Opportunities can be much more valuable than advice. Think about who you're sponsoring and delegating to. Share the spotlight in your team.

- Plan to give away your job.

9

What's Next?

We started this book with a journey to understand what your job is. Since then, you've unpacked your scope and primary focus, mapped your org, developed strategy and vision, prioritized work, led projects, navigated obstacles, modeled good engineering, and brought your colleagues up a level. It's been a journey! And now we're at the final topic: we're back to *you*. But instead of looking more at what you're doing *now*, we're going to look at where you go from here. We're going to look at leveling up *yourself*.

What does leveling up even mean? It depends on the circumstances and it depends on you. So we'll start by returning to a theme throughout this book: *what's important?* We'll look at the big picture of your career, where you're going, and what you need from your next steps. Then we'll take a look at your current role and evaluate whether it's a step toward where you want to go.

The staff engineer role is loosely defined, so it's not surprising that the paths onward are loosely defined too. We'll look at some of the options you have, and I'll share stories from other people who have traveled onward from staff+ roles. These stories are just a sampler of the range of things you can do, but they might serve as inspiration for your own journey.

To finish, we'll consider the influence you'll have throughout your career. As a senior person, you're one of the leaders of our industry. You're responsible for the choices you make, and you'll influence other people's choices too. And you're the only person who can drive your career. We'll start there.

Your Career

A friend at a huge company once compared his career journey to playing *Diablo*, a classic role-playing video game. "I fight all the monsters and clear the dungeon," he said, "and eventually I collect enough experience points to go up a level. But then...I just start again in a new dungeon and the monsters have *also* gone up a level! What's the point?!"

What *is* the point, for you? Where are you going?

Back in Chapter 2, we drew three maps to describe your work. Let's draw a fourth now: the trail map. Imagine your career as a journey across mountainous terrain. There are many paths marked on your map, some well-traveled and some overgrown. Some trails have limited visibility, but you can stay oriented by catching glimpses of landmarks as you travel. Some are twisty and you might need to take a path that seems to lead *away* from your destination, but that's the only way to where you want to go.

Not all destinations are on the map, and many of the interesting ones can only be reached by leaving the trail. If you always choose your next destination based only on where the marked paths lead (see Figure 9-1), you might just keep following other people's footsteps to the next dungeon and miss out on places you actually wanted to go.

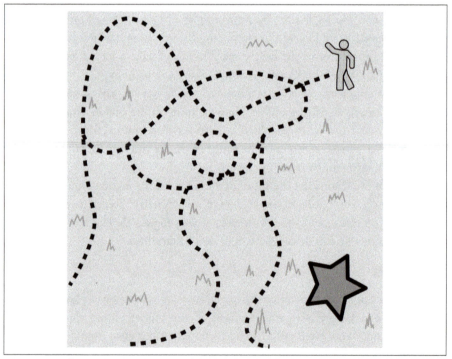

Figure 9-1. Following local trails. None of the marked paths go to the destination, but if you have a map you'll know the right place to leave the trail.

But where are you going? Maybe you've got a clear destination with milestones along the way. Maybe you don't know *exactly* where you're going, but you know the rough direction you want to travel. Or maybe you're not going anywhere in particular: you're just enjoying the journey.

For the rest of this chapter, I'll assume you want to go *somewhere* from here. Career progression is often framed as climbing a career ladder and growing your seniority, responsibility, power, and wealth. That's just a subset of the trails, though. Let's look at what's important to you.

WHAT'S IMPORTANT TO YOU?

Staff engineer Cian Synnott has written about creating a priority list (*https://oreil.ly/4ShIX*) as a way to stay oriented and make sure his work is supporting what he wants from his life. Creating a list like this is a great way of introspecting about what matters to you. What are your career and life priorities? Here are some common ones:

Being financially secure
It might be most important to you to pay off debts, prepare for college fees, or save for your retirement.

Supporting your family
If you've got family who depend on you, you might accept a job you enjoy less for the sake of a salary that lets you take care of them. You could be optimizing for a steady paycheck with good benefits and no fear of layoffs.

Having a flexible schedule
You might want flexibility, like a schedule that accommodates childcare or eldercare or a chronic illness or disability. Or maybe you just want a schedule that gives you lots of free time to do the things you enjoy.

Learning a lot
Maybe you want the intellectual satisfaction of becoming world class in a particular domain or being the kind of generalist who can step up to any challenge—maybe you're the person we talked about in Chapter 3 who saves the Starship *Enterprise*!

Being visible
It might be important to you to have the respect and admiration of your peers, to be "industry famous," or to be *representation* in a prestigious role,

someone that other people can see being successful so that they can imagine it too.

Doing cool things

You might want to work on cool and exciting projects, things you find energizing and fun.

Challenging yourself

It can feel amazing to look back at challenges you tackled that were bigger than you could have imagined.

Building wealth

Can we take a moment to appreciate how *incredibly lucky* we are to be in an industry that is (currently) highly compensated? You might be looking to get as much money in the bank as possible over your career.

Working for yourself

Maybe you don't enjoy having other people make decisions that affect you, or just want the experience of being your own boss. If you ultimately want to set up a company or work independently, you might be trying to build the skills, experience, and contacts to feel confident doing that.

Making a difference

Maybe your lifelong goal is to make the world better, to leave a legacy that outlasts you. That could mean teaching, inventing something, or building communities that make tech a more friendly place. Or it might mean using technology as a tool to cause the real change you want: creating products that improve people's lives, or being able to support causes that need you.

Enabling your vocation

You might be working to support the thing you *really* care about: succeeding in a music career, say, or making your hobby farm viable.

There are lots of other things you could optimize for: making friends, traveling the world, taking care of your health, and so on. There's no right answer: it's personal to you, and it's likely to change over your career. Take a moment to think about what's on your priority list at this stage of your life.

WHERE ARE YOU GOING?

Your priority list will keep you oriented, but it doesn't tell you exactly where to go. For that, you need to draw your trail map and mark in some milestones. The amount of detail in your trail map and the span of time it covers will be up to you, but if you've got a faraway goal, plot some steps that will take you closer. Where do you see yourself in five years? And what does that mean you need to do now?

Like the other maps, the trail map will be better if you don't draw it alone. If you base it only on your own experiences, you won't be able to find the nonobvious paths, and the trails will be limited to what you can see from where you stand. You can expand your perspective by reading, attending conferences, and especially by asking other people about their journeys. Seek out the people who have taken paths you're interested in and talk with them.

Your manager may be able to help with your career, but that's never guaranteed, and especially now. At staff+ levels, your manager might not even know how to help you: you're likely on a path that they haven't taken. Being senior doesn't mean you know it all, though, and you'll still need help and guidance. That means it will be up to you to seek out other sources of advice, teaching, and guardrails, and to look for the opportunities that will help you grow.

If you do discover an ambitious path you're excited by, really consider it. It can be tempting to choose your route based on what's most clearly signposted, but think beyond what you can see from here. Yes, some paths will be harder, feel riskier, or have less available support, but if there's something ambitious you wish for, don't limit yourself. If the role, impact, or lifestyle you want ends up not feeling achievable from where you are, you can still take steps toward it, maybe moving to a vantage point that will tell you more about where to go after that. A career is a long time.

WHAT DO YOU NEED TO INVEST IN?

If you imagine yourself as having succeeded at your goals and achieved whatever is important to you, what does that look like? What skills does that successful future person have that you don't currently have? What lines are on their résumé? How did they spend their time and who did they get to know? Ask yourself these questions as you decide what roles and opportunities to take.

Building skills

I've sometimes heard people say they're really *not good at* something: functional programming, say, or telling a good story. I prefer a different framing: that's not a skill you've leveled up (yet). If you'll forgive one final tortured video game metaphor, think about games like *Final Fantasy,* where every mission gives you ability points to improve your skills. Let's say you get new ability points this year, and you currently have 14 points in wielding Python (Figure 9-2). You can invest some points in getting to level 15, or you can ignore that skill and focus on other things, like getting better at JavaScript. Or maybe you'll get your first points in something entirely new.

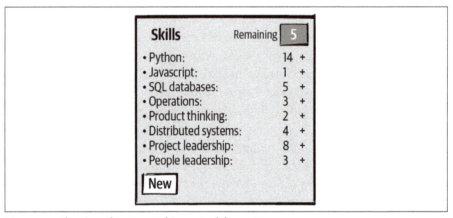

Figure 9-2. Choosing where to put this year's ability points.

I've often heard people say that you should focus on what you're good at. I don't entirely agree: you should put points into what you *want to* be good at so you can build up those skills.

Everything in tech is learnable, if it's worth the time investment. It might not be! There are more available technical domains and skills than there is time to learn them, so you can't ever be an expert in everything. That's OK: strong teams are built from people who each have a subset of the necessary skills. But if there are skills you need to get you to your goal, or gaps in your abilities that make you feel insecure, assume they're just unknown, not unknowable. You just haven't put ability points there yet.

That said, don't make your journey harder than it needs to be. If the work you're doing all day fills you with dread or exhausts you instead of exciting you, look for a different path to your goal.[1] Charity Majors, CTO of Honeycomb, points out that (*https://oreil.ly/jumRI*) keeping up with our fast-paced industry

---

[1] I love Dina Levitan's story (*https://oreil.ly/lt25t*) of realizing that she wasn't *bad at* axe throwing, she was just using a kind of axe that didn't work for her throwing style. As she says, "We can all learn to hit the target... but it's important to choose the right axe."

means managing energy: "If you want a sustainable career in tech, you are going to need to keep learning your whole life... Make sure that you a) know yourself and what makes you happy, b) spend your time mostly in alignment with that. Doing things that make you happy gives you energy. Doing things that drain you are antithetical to your success."

Practicing skills you find difficult may turn them into strengths and reduce how much energy they cost—or they may never stop being a slog. Decide what the skill is worth to you. Each of us will find different things easy, and where to spend your points is up to you.

Imposter Syndrome

New grads just entering the industry would be stunned to see how many of the staff engineers they look up to feel like imposters. Depending on the path you've taken, you might feel insecure about, say, technical strategy, or influencing people, or systems design. Some systems folks feel weird about not having logged more hours writing code. Some product engineers suspect that they're "supposed to be" more adept at operations.

Imposter syndrome is a horrible, insecure feeling: it can even make you do less good work because you don't feel safe taking calculated risks. Take comfort in the fact that it's common, even at this level. Find opportunities to put ability points into something that makes you nervous, and show yourself that it's just another learnable, knowable topic. But most of all: if you're "impostoring," try to give yourself a break. Nobody knows everything.

Building a network

Skills are rarely enough to take you where you want to go. You need contacts too. Depending on the business article you read, you'll hear that upward of 70%, maybe as many as 85%, of jobs are not actually published: they're found through networking. When an internal project needs a lead, there'll be a buzz of back-channel conversation among the nearby folks in leadership roles to see who they recommend. When a conference needs a speaker or a project needs a paid consultant, the participants will reach out to people they know. It pays to be known.

It pays to know people, too. Having contacts in various roles gives you insight into how people in those roles behave, what "competent" and "professional" look like for them, and how they communicate.[2] Having a strong network means you'll always have experts to go to for advice and to learn from.

CTO Yvette Pasqua has spoken about how to build a network in a sustainable way (*https://oreil.ly/JHTrC*), without burning all of your introvert energy. It doesn't matter if networking isn't innate or comfortable for you, she says: "If you don't know who to talk to or how to start, a little secret is that none of us do." Pasqua reaches out to people she'd like to talk with and invites them to chat about a specific topic she thinks they'll find interesting too. She recommends joining groups and communities—but only the kinds that give you energy—and connecting one-on-one with people at events.

That last one can be excruciating for introverts and socially awkward folks, but there are a ton of mechanical tricks you can learn to make it OK. I love author and "recovering awkward person" Vanessa Van Edwards's Science of People (*https://oreil.ly/fkXsx*) website for learning some of this "humaning" magic that other people seem to have been born knowing how to do. Check out her article "How to Network" (*https://oreil.ly/JXjuH*), for example, for tips on talking to people at events like where to stand, how to remember people's names, and what to talk about. All of this stuff is learnable.[3]

Building visibility

If you have skills, but nobody knows about them, you won't get invited to use them. Let people see you solving problems, asking insightful questions, or showing up with a clear strategy when there's chaos, and they're more likely to think of you when an opportunity arises. You'll meet interesting people too: they're more likely to reach out because they see you're working on something that's relevant to both of you.

You may choose to build an external reputation, making a name for yourself with open source contributions, industry working groups, articles, podcasts, videos, conference talks, and so on. These kinds of ventures are usually optional, but can be incredibly helpful if you're looking for roles or connections, and they're ways you can have good influence across the industry too. Some employers

---

2 This is another reason representation matters so much.

3 Half of the "confident" people you talk to at a conference are running the interaction in software and hoping they're Doing Social properly. I've, uh, heard.

encourage external contributions, either to help with recruitment or to draw attention to a company's product or service. If you take this path, expect being a "public person" to take some investment—this is one of the places you'll be spending your ability points.

Being offered opportunities doesn't mean you need to take them, but if you do get one you want, don't waste it: show what you can do. If you have a proposal accepted for a conference, for example, don't present a talk you threw together on the plane.[4] If you've joined an open source community, don't start out by picking fights. Let people see you do the work with grace and aplomb—be visibly competent.

Choosing roles and projects deliberately

The most time-efficient way to build skills, visibility, and contacts is as part of your job. You'll get better at whatever you spend time on. In fact, it's easy to gain a specialty *accidentally* just because it's what you're doing at work: one experience leads to another, and next thing you know you have a specialization. Spend five years writing storage systems, for example, and you're going to get good at writing storage systems: you'll have relevant skills, you'll work with other storage experts, and you'll have storage-related lines on your résumé. When an ex-colleague is hiring a storage expert, they'll think of you. Spend the same five years on a popular mobile app, and you'll build a completely different set of credentials. It's easy to get typecast.

So choose roles that will give you the experiences you want to have. There are some things you can only learn at big companies, others you can only learn at small companies. Some things will be easier as a manager, others only if you're really hands-on. If you're not sure what you need, find someone who's doing the role or living the lifestyle you want, and ask them what key experiences brought them to where they are today. (You can sometimes snoop their résumé on LinkedIn instead, but real-life conversations will tell you more.)

You get better at what you spend time on, so be deliberate about choosing roles and projects that will give you skills you want to have. Mason Jones, who has been an engineering leader at over 10 startups since starting with Travelocity in 1995, agrees: "Consistently and mindfully taking positions where I could

---

4 If you're one of the tiny number of people who can throw together some slides, stroll on stage and deliver an unrehearsed talk that gets a standing ovation, I'm not talking about you. Keep doing your wizardry. Tell me your secrets.

expand my knowledge and broaden my experience has been the single most valuable thing I've done throughout my career."

Your Current Role

Every job should help you grow toward your long-term goals and meet your immediate needs. Unfortunately, often people end up in roles that don't do either. We'll start this section by looking at whether your job is good for you, then move on to evaluating whether it's possible for it to match your wish list.

FIVE METRICS TO KEEP AN EYE ON

Is your current role taking you closer to your goals? Might it be doing the opposite? Experienced engineering director Cate Huston offers five metrics (*https://oreil.ly/7JLMI*) for evaluating your job health:

- Whether you're learning
- Whether you're investing in transferable skills or navigating dysfunction
- How you feel about recruiting other people to your team
- How confident you feel
- How stressed you feel

A great job situation keeps you growing toward your goals, and your self-confidence and abilities stay high. A bad one—which has stagnation, working with a bully, lack of support, impossible deadlines, or other difficulties—might get worse slowly enough that you don't notice when you're well past the point where you should have walked away from it. I've seen friends in unhealthy work situations become convinced that they don't have the skills to get hired somewhere else. As a result, they stay in roles where they've stagnated, and the lack of skills becomes self-fulfilling. As Huston says, "Sometimes five years of experience is just...the same year of experience, five times over." Oof.

In another article, Huston explains that (*https://oreil.ly/qw2J2*) while your employer is buying your time, they're only renting your "brand." OK, the notion of having a personal brand will feel squicky and artificial to a lot of engineers, but think beyond polished people with expensive hair and expensive fonts: your brand is how you're perceived by other people. If your job is making you less employable, Huston says, "I hope your employer is paying a lot of rent—because they are destroying the market value. At times that might be worthwhile, but often it is not, and people realize that too late."

We all have good weeks and bad weeks, so one model I've recommended to friends is to track these metrics over a few months (see Table 9-1) and see how things are trending over time.

Table 9-1. Tracking the signals of job health described in Cate Huston's "5 signs it's time to quit your job" (https://oreil.ly/7JLMI).

Signals	Are you learning? Are you growing?	Are you learning transferable skills or just how to cope with your org's dysfunction?	How do you feel about recruiting friends to your company?	How's your confidence and how capable do you feel?	Is your job physically good for you?
Scale	0: stagnant 5:rocketship growth	0: learning to cope in this org 5: learning transferable skills	0: morally conflicted 5: wildly enthusiastic	0: confidence being eroded 5: confidence growing	0: stress stress stress 5: feeling healthy
<date>					
<date>					
...					

Tracking metrics can shield you from recency bias and let you look at a bigger picture over time. If you can look back and see that things have mostly been good, you'll be less likely to rage-quit over a bad month. But if you notice that you *keep* having bad months, or that your metrics are trending worse over time, notice that you might be in a situation that's not good for you. Consider Captain Awkward's (https://captainawkward.com) *Sheelzebub principle*, a question to ask yourself about bad relationships: "If things stayed exactly like they are, would you stay: Another month? Another 6 months? Another year? Another 5 years? How long?"

CAN YOU GET WHAT YOU WANT FROM YOUR ROLE?

Is your job moving you toward your long-term goals? Is it a healthy environment? Take a moment to look through your priority list and evaluate how well your job is meeting your needs (see Figure 9-3). Make sure you appreciate what's great, as well as identify what's missing. It's easy to take the good stuff for granted.

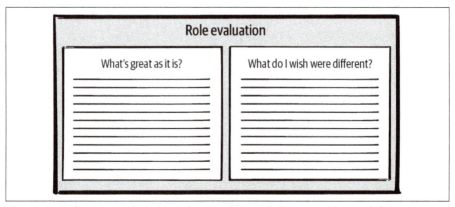

Figure 9-3. Evaluate your role and see what's working and what you wish were different.

No matter how great your job is, you almost certainly didn't conclude that it's perfect! You'll rarely find a role that is the best for *everything* on your priority list. Making the most money is often in tension with doing good in the world. Learning the most can mean not taking on the most prestigious roles. Big career accomplishments can mean not having time and energy left for other aspects of your life. This is why nobody else can make career decisions for you: we'll all make different trade-offs. But look at what's not optimal and think about whether it's something you can change without compromising too much somewhere else

SHOULD YOU CHANGE JOBS?

If you want to fix something that's not working, you have two options: modify your existing job or move to a new one. Let's look at some of the reasons to do each.

Reasons to stay in the same role or company

If your current role is giving you most of what you need and taking you where you want to go, it can be rewarding to continue doing the same thing for a long time. Staff engineering benefits from the longevity, domain knowledge, and relationships that you build over time in one place. Here are some other reasons it's good to spend a long time in one place:

Feedback loops

Staying in one place for longer gives you the feedback loop that comes from seeing the consequences of your actions. When engineers move around a lot, everyone's seeing the results of someone else's past decisions

instead of the outcomes of their own. You may also get to see the colleagues you leveled up become senior or staff engineers and then become role
models themselves.

Depth

The more you know a single domain or a single stack, the deeper and more
nuanced your understanding will get. It takes time to intuitively understand something so well that you can build on the knowledge. It's also
faster to do things you've done before; you'll be able to make progress more
quickly.

Relationships

You've invested time in knowing people all over the organization, and you
have people you trust and enjoy working with. You've built up enough
mutual goodwill that even the biggest technical disagreements are collegial,
not heated. That's an asset that takes time to build up again.

Context

After investing time and effort into learning how to navigate your organization, you have a skill set that might not translate to another one. You've
figured out the OKR process, you know the shadow org chart, and you
know how to get things done.

Familiarity

You know the work, the schedule, and the people. If you observe particular
religious holidays, pick your kid up from school every afternoon, or you
always play bocce at lunchtime on Thursdays, you've already set up your
schedule to make that happen. It just works, and you're reluctant to change
anything.

Reasons to move

But there are also good reasons you might want to move around at intervals:

Employability

If you stay at one place for a very long time, you might be learning how to
work in that culture rather than learning transferable skills. The world outside can shift, and you can get left behind. Keeping more skills and
domains fresh can keep more doors open.

Experiences

There will be a limited number of experiences available in any one place, and a limited number of people to learn from. Once you've collected everything available, you might be ready for something new.

Growth

It can sometimes be easier to get a step up in level or scope by changing jobs. Maybe the next level feels too far out of reach to be realistic, or involves the kind of politics or work you're just not interested in. If you're struggling to get your name in the ring for the important, challenging, or visible projects where you are, it can be easier to find a new job than get a promotion you're hoping for.

Money

Changing jobs can be the fast track to higher salaries. While some companies stay up to date with their current employees, often new hires can negotiate better salaries, stock grants, and hiring bonuses.

Mismatch

Not all paths to growth exist at all companies. If you're looking to become an industry expert on a topic your organization doesn't really need an expert in, if the projects you're energized by aren't the ones that your leadership wants to invest in, or if there are more senior people than there are leadership opportunities, it might be time to move on. Not all roles are available in all places.

The right next steps will depend on what you need. In the next section, we'll look at some of the paths onward from here, some staying where you are, others changing roles or companies.

Paths from Here

Where do you go from here? Let's look at a sample of your options:

KEEP DOING WHAT YOU'RE DOING

If your job is giving you what you need, there's no need to change anything. I want to emphasize that because our industry puts a lot of focus on changing jobs frequently, and the regular "new job" announcements can make you feel like you should be moving too. If you're in a growing niche, you might be able to stay in the same team and still have plenty of room to learn and grow for decades. Or you might not be looking for further growth at all: you may just want to use your current skills and keep doing much the same job until you retire.

If that second one is you, be a little wary about industry changes making your skills obsolete. Technologies and your business will change, and even leadership skills can slowly become old fashioned if you don't keep them up to date. Social norms, communication styles, best practices...it's all going to change. So staying still probably means moving and learning a little, just enough to stay up to date.

WORK TOWARD PROMOTION

Staying in the same role can often be a path to the next level up. As your influence, knowledge, and impact expand, you and your manager may start to feel that it's time for you to be promoted.

Progressing to higher levels can be lucrative and can give you the credibility that comes with a more senior job title; as I said in Chapter 1, it saves you having to spend time and energy on proving you should be in some conversations. And, honestly, being promoted or being offered a bigger job just *feels nice*. But unpack that feeling: is it really about the level, or are you actually looking for cash, prestige, interesting challenges, broader scope, respect, getting in "the room," a sense of progression, or something else entirely? It's fine to want any of these things, but be sure that the next level is going to give you what you want, or you may end up quickly feeling underwhelmed by your new title.

Understand what promotion means at your company: does your director decide who gets promoted? Is there a promotion committee who will review your work? There may be a written career ladder with expectations at the next level, but these expectations are often low on detail about what's really expected. There may also be restrictions on how many people can be promoted, or how many people can exist in a particular role: you might not be able to get promoted unless there's an appropriately sized scope or project that needs leadership.

If you're looking for promotion, discuss it with your manager and ask for their guidance. Connect with other people at the next level and understand what their path was. Looking for footsteps to follow can be intimidating if the other person has been in the role for a few years. Remember, you're looking to be as impactful as they were when they got promoted, not where they are now.

WORK LESS

Success can mean working less in your current role. One engineer I spoke with, Jens Rantil, swapped a staff engineering role for 80% time and a 20% pay cut at a much smaller company. As he said, "Every Thursday is a Friday! It's amazing!" Rantil observes that moving to 80% time is the first time many people set a price tag on their free time and decide what having more of it is worth. (Remember that a pay cut goes beyond immediate salary: it's also likely to affect retirement savings.) While 80% is the most common schedule change, I've seen engineers arrange to work 60%, 40%, and even 20% time; some employers may be interested in retaining your skills for one day a week.

If you're cutting back your hours, be deliberate about where those hours are coming from. It might be easiest to drop your focus time and just go to meetings, but that may not be what makes you happy, and it might mean you don't achieve the outcomes your manager wants either. Or you might find that you end up working extra unpaid hours just so you can do the fun part of your job: if you weren't able to avoid working overtime at five days a week, be really clear about why you think you can stick to working four.

Make sure to align your expectations with your manager's and team's. If you still want to work toward a promotion, or be considered to lead interesting projects, make sure your manager knows. Agree on how you'll handle specific situations like on-call, holiday weeks, or weeks where you're out sick on one of the other days. Be warned that many teams won't be enthusiastic about having less of your time; from the headcount allocation point of view, the team may have whatever number of engineers it has, and if you work fewer hours, they don't get someone else to make up the slack. But you may be able to achieve a lot in less time. One person I spoke with said that they don't get much less done since they started working five-hour days—they only have four hours of productive work in them on any given day anyway.

CHANGE TEAMS

If you're ready for a change but happy with your current employer, an internal transfer can be a great move. You keep a good amount of your context, relationships, credibility, and social capital, but you get to start fresh on something new. Burin Asavesna, a software architecture lead at Hilti, told me that he thinks of this kind of restart as being like an experienced player of a game making a new low-level character: technically you're starting from scratch, but in reality you already know how the game works and you'll fly up the levels.

Moving between teams or organizations can be an excellent way of building bridges: you'll still have contacts on your own team, and you'll bring knowledge and culture with you to the new one. You'll also bring perspective: you already have an outside view of the team and how they're perceived by the rest of the organization.

BUILD A NEW SPECIALTY

The breadth of the tech world means that there's always something new to learn. You might enjoy learning about something very adjacent to what you already know, adding a new dimension to your knowledge, or begin putting a lot of ability points into something entirely new. You might even build a new accidental specialization. A lot of the most interesting innovations come from people who are very comfortable with more than one domain: interesting things happen on the boundaries!

Building a new specialty might mean more than a new team; it might mean moving to a different career track, either temporarily or permanently. Former principal engineer Lou Bichard has written (*https://oreil.ly/51xAs*) about moving from being a "product-minded engineer" to officially becoming a product manager. As he says: "Taking a little time out to do something a bit different might help you bring some new perspectives into your work."

EXPLORE

Some companies will let you take short-term gigs, embed in a team, take part in a rotation program, or just try out a new team for a while. One staff software engineer at a huge company told me about doing this kind of exploration after being on the same team for six years. Her company had a wide range of opportunities, including rotation programs, so she decided to take some time to explore what was out there. Over two years, she tried out three teams: a large site reliability team working on mature infrastructure, a small research team working on a recently released product, and a medium-size team working with a nonprofit to create a new open source product. After these vastly different experiences, it was clear to her that the research team matched her interests and had lots of growth opportunities, and she's been there for the last year and a half.

TAKE A MANAGEMENT ROLE

Are you feeling a pull toward management? Some staff engineers move entirely onto the management track and continue to grow there. Others take a stint in management before returning to an IC role.

In a famous article, Charity Majors introduced what she calls the engineer/manager pendulum (*https://oreil.ly/1eBJs*), the idea of deliberately moving back and forth between manager and IC roles every few years. Majors rejects the idea that you have to choose a lane and stay there:

> The best frontline eng managers in the world are the ones that are never more than 2-3 years removed from hands-on work, full time down in the trenches. The best individual contributors are the ones who have done time in management. And the best technical leaders in the world are often the ones who do both. Back and forth. Like a pendulum.

Majors emphasizes that management should never be seen as a promotion —it's a change of profession with a different set of skills to learn. There should be no change in status when you go from people leadership to technical leadership or vice versa: each will build a separate set of skills, and will enhance the skills on the other side. But she doesn't recommend trying to do both at once: "You can only really improve at one of these things at a time: engineering or management."

Will Larson argues (*https://oreil.ly/wnP3C*) that a hybrid engineer/manager role is not *always* a bad choice, so long as you've already built up solid experience as both team manager and technical contributor. But he agrees that if you're trying to learn either set of skills on the job, you're going to have a hard time: "If you've built up your experience as both team manager and technical contributor, then sure give it a whirl if it's what checks the most career boxes for you, but I do consistently recommend against folks starting their management career in such a role."

If you're going into this kind of hybrid role, have a plan for how you're going to make it sustainable, perhaps by having a bench of other senior people you can delegate to or lean on when you need them.

TAKE ON REPORTS FOR THE FIRST TIME

What if you haven't tried people management before? If you've never been a manager or had direct reports, it can be intimidating to take on your first management role in your later career. But here are three reasons you might be ready for your first direct reports:

- If a future goal needs you to have management experience, you'll eventually need to start building that skill set.

- If you're at a company or on a team where decision making and context only come to folks on the manager track, you might decide that the management track will give you more leverage to get projects done.

- Or if you're at the top of your career ladder and are interested in the business problems of the next level up (and can't convince your organization to add another rung), managing a team may be the next step to growth for you.

Some companies have the concept of staff engineers with reports, some don't. Taking on reports might require you to change tracks.

If you're used to being an IC at the level of a senior manager, director, or VP, it might be tempting to argue that you should manage an organization at the same scope. Amanda Walker, an engineering director in security at Google, and a past staff engineer with reports, advises against this; she recommends that you spend some time as a line manager before taking on a more senior organizational role: "Just as having been a software engineer helps me be better at managing software engineers, having been a line manager helps me be better at managing other managers. It's easier to coach a sport you have played well."

If you've been used to working at an organization-wide scope, though, it may not be an enticing prospect to go back to managing sprints for a single feature team. One possible compromise is to look for an opportunity to take a tech lead or team lead role for a cross-team project, taking on management responsibility for a small number of people on that team.

How you show up as a manager affects the lives of your reports, so if you're going to be a manager, invest in it. Majors says that (*https://oreil.ly/7xGlc*) your minimum tour of duty should be two years:

> *If you really want to try being a manager, and the opportunity presents itself, do it! But only if you are prepared to fully commit to a two year long experiment...It takes more than one year to learn management skills and wire up your brain to like it. If you are waffling over the two year commitment, maybe now is not the time. Switching managers too frequently is*

disruptive to the team, and it's not fair to make them report to someone who would rather be doing something else or isn't trying their ass off.

Committing to management means accepting that it will take time. Expect not to do nearly as much coding, architecting, or technical work as you would otherwise—and understand that you might not get to do any. In *The Manager's Path*, Camille Fournier writes, "It's OK to feel nostalgia for the simpler times, and a little bit of fear for what you're giving up. But you can't do everything all at once. Becoming a great manager requires you to focus on the skills of management, and that requires giving up some of your technical focus."

Majors agrees (*https://oreil.ly/gs701*): "If you're a manager, your job is to get better at management. Don't try to cling to your former glory."

FIND OR INVENT YOUR OWN NICHE

Senior leadership roles often have specific needs. Even at the same company, one staff engineer opening will need someone with strong architectural skills; another will want a skilled project leader who is great at crossing organizations; a third will be looking for extra leadership bandwidth. The more senior you get, the more likely you are to be looking for a role that needs *your specific skills*, rather than shaping yourself to fit a generic role.

Molly Graham (of the "give away your legos" article I mentioned last chapter!) says that careers come in two phases (*https://oreil.ly/XhUGB*): first learning what your strengths are, and then finding "holes that are shaped like you." "Happiness," Graham says, "is going to come from finding roles that fall in the intersection of what you love doing and what you are great at." But she adds:

Beware the role that sounds absolutely tailor made for you but also feels completely exhausting when you imagine doing it. Doubly beware if the job is "fancy"—where your friends and family are going to think it's cool—because then your ego gets in the mix and wants you to take it even though your gut says that you will hate most days on that job. That venn diagram—things you're exceptional at but hate doing—is one that can lead to career mistakes.[5]

---

5 Graham adds, "I've found that people that know you well are always going to be the ones that find you the phase 2 roles that are 'shaped like you.' People that don't know you are always going to offer you the job you just had." This has been *exactly* my experience too.

One way you can find a role shaped like you is by carving it out yourself. If you get this opportunity, you can fill a gap that your organization has and create a job you love at the same time. When Keavy McMinn was ready for a change at her current company, she found it liberating to openly discuss her goals with her manager: "This is what I'm good at and really want to do. How can I be of most value to you and the company?" Together, they crafted a mutually beneficial new role as the technical adviser to the CPO at Stripe.

McMinn says that creating her own role was feasible because she was in a privileged position and was comfortable with the risk of asking to do something different. This path won't be available to everyone. But, she says, it's surprisingly common: "It might be helpful for you to know that *people do this*! Give yourself permission to explore the idea of crafting a new role, together with the people who can support you. No one else really can or will drive that conversation for you. See it as an experiment, even—which can relieve some pressure!"

Another great piece of advice I got, from leadership coach Fabianna Tassini of Confidantist, is that if you get a chance to design your role, you should include a lot of what you love to do. Make the work that gives you energy about 70% of your job. The other 30% should be things you'd like to practice and get better at. (Of course, the company needs to actually need the role you're creating; you may need to compromise to find something that fits you and your employer.)

DO THE SAME JOB FOR A DIFFERENT EMPLOYER

Starting in a new workplace can give you a completely different perspective and fill experience gaps that you have. Depending on the experiences you're looking for, you might choose a different organization size, technology stack, domain, or culture. Just try not to "rebound," if you can. Molly Graham writes (*https://oreil.ly/PQz9Q*): "Sometimes when you're in a job, particularly if you're not happy or burnt out, you have a tendency to pick a 'rebound job.' Just like a rebound relationship, a rebound job just helps you get out of your current situation, but it often isn't the best or healthiest long-term choice." She adds, "Picking the opposite of what is currently making you miserable won't lead to happiness, it just helps you get out of a bad situation."

If you have the option, take your time about choosing your new role and understand what you're actually looking for. Don't just jump for the first good-enough recruiter mail that crosses your path. You deserve better.

Since staff+ roles still mean wildly different things in different places, have an explicit conversation with future employers about what it means to them and what your job would be. As staff engineer Amy Unger writes (*https://oreil.ly/ o2zWd*), "It's likely that each company and even each manager you talk to will have assumptions about what combination of skills they're hiring for and an inability to articulate them." Ask a lot of questions.

Staff+ interview slates are far from standardized: you might get asked about coding puzzles, systems design, previous projects, what you'd do in various leadership scenarios, or "Tell me about a time you..." Many organizations will share the interview slate in advance. If they don't, it's fine to ask your recruiter what to expect, so you can make sure you're prepared. Use the questions you get as a hint about how the new company sees the role, and remember that you're interviewing your interviewers too.

CHANGE EMPLOYERS AND GO UP A LEVEL

Changing companies can be an opportunity to reinvent yourself. It can also allow you to find roles that just aren't available where you are. If you're struggling to prove yourself through the promotion process on your current ladder, it can sometimes be easier to interview for the next level elsewhere. If your company is growing too slowly to need more senior leaders, or doesn't have room for another principal engineer or senior manager until someone quits, you can sometimes find more opportunity by leaving.

Changing companies and going up a level is usually easier if you're already an expert in the new company's domain. It's a harder sell to say that you want to operate at the next level up and *also* learn about, say, the health care or construction industry.

If you're looking at a role that's the next level up, be careful that the industry really does regard it as an increase. The *levels.fyi* site can help you calibrate what different titles mean in different companies.

CHANGE EMPLOYERS AND GO DOWN A LEVEL

Sometimes you'll take a path that seems to go backward in some aspects: a smaller scope, less money, a less prestigious job title, or a role where you're a beginner. If you think about your path only in terms of *everything* improving or increasing, you're going to limit your options. In particular, moving to a bigger company will often come with higher expectations and corresponding downleveling. Going down a level can often be possible without decreasing your compensation, and it can let you shore up your technical foundations, or do more of a kind of work you enjoy. Josh Kaderlan, now a senior engineer at Datadog, told me why he was comfortable giving up the staff title that he had at his previous company: "If you make title a gating factor for a new job, you increasingly limit your opportunities. Being in this new environment has been rewarding, especially given that I am no longer always the most senior person in every conversation, and I have the opportunity to learn from people who have more and different experience than I do."

Another engineer, Stacey Gammon, told me she thinks of her career as being similar to the engineer/manager pendulum, but moving between technical leadership and hands-on coding roles. When we talked, she was leaving a leadership-focused principal engineer role at a publicly listed company and weighing offers from several other companies for much "smaller" roles where she would have more time to code.

SET UP YOUR OWN STARTUP

If you're really looking for a change, and want to be your own boss, you might consider the challenge of setting up your own company. James Kirk, a former staff machine learning engineer at Spotify, told me about leaving his role and cofounding a startup as CTO. He told me: "I was interested in starting something myself because it seemed challenging and rewarding, and that's the kind of itch that just gets worse over the years until you scratch it. I started connecting with some local VCs and their communities a couple years ago and through them met the person who is now my cofounder. We starting throwing ideas around and eventually found some things that we were really excited about and that's when we took some VC cash, quit our day jobs, and started in earnest."

If you're preparing to set up your own business, that can be a good reason to focus on compensation for a while first: you're building up a safety net of cash to give yourself some time without much income. Kirk adds: "I don't think that, practically, I would have been comfortable taking the risk if I didn't have a few years of healthy tech comp stashed away before leaving."

GO INDEPENDENT

Another version of going to work for yourself is working independently, through consulting, contracting, indie app development, training, or other part-time work.

Emily Bache, a software consultant and author of several programming books, says that the real benefit of being independent is the freedom: "I have a lot of control over how I spend my days, I have lots of time to read and learn things and share my ideas. I get to go to interesting places, meet interesting people, and work on interesting coding problems."

Bache emphasized that working independently benefits from a strong preexisting network, so potential collaborators and customers will know what you can do and will approach you. It helps to keep your public presence up to date too: "Marketing is a constant—I aim to speak at about 10 conferences and local events each year, plus publishing articles. I also invest time posting on social media—Twitter and LinkedIn—so that people can find me through them."

Consulting is not for everyone. One consultant, Vlad Ionescu, warns: "Even for folks that want to take it up mid-to-late-career, it's a big switch (different skills are needed, finding clients, etc.) and it's not as glamorous as many find it (usually less money than a stable FAANG job, more stress, etc.). A lot enjoy it and are a fit for it, but a lot of wide-eyed hopeful engineers get badly burned."

So understand what you're getting into. Make sure you're clear about the trade-offs, and accept them. And just like with setting up a startup, you can reduce your risk by starting with a safety net of cash.

Finally, remember that being independent means you'll be missing many of the support structures you might have expected in other roles. Chris Vasselli, who left a staff engineering role at Box to become a full-time indie app developer, advises: "While you're at a company, learn from as many teams/experts as you can. Frontend, backend, desktop, mobile, design, security, QA, localization, build & release, even (especially!) marketing, growth, and biz dev. When you're an indie, you are responsible for all of these." Be prepared to wear a lot of hats.

CHANGE CAREERS

After years in technical roles, some people feel a pull toward doing something different. That might mean a career shift into education, academia, policy, or research, bringing your technical experience and background to solving a different kind of problem.

My favorite example of branching out in a new direction is Peter Lyons, who left his role as a staff engineer at Intuit, teamed up with his partner, chef Christella Kay, and set up a retreat for programmers (*https://oreil.ly/aJtsl*) in the Adirondack mountains. Now he slings pancakes instead of code. He says that the COVID-19 quarantine helped him realize what was important to him: "It motivated us to make big changes to flip our lifestyle so that we could spend our days working together on something we cared deeply about."

Prepare to Reset

If you do move to another job, be prepared to be a beginner again. All of the maps you drew in Chapter 2 are now out of date! Even if you're in a new role in the same organization, you'll likely be starting from a different place and different context. You'll need to build perspective again and draw a new locator map. You'll need to learn the terrain, culture, and politics and draw a new topographical map. You'll need a new treasure map to help you understand where you're going.

Author and distributed systems engineer Cindy Sridharan warns against (*https://oreil.ly/HfjSK*) trying to do a new job using the rules of the previous one.

> *Not all newly hired senior leaders are entirely committed to or feel comfortable turning themselves into the leader the organization truly needs, rather than the leader they've grown to be over the past years. Many leaders take the opposite approach of trying to mold the organization in their image or the image of the past workplace. Engineering leaders brought into embattled organizations tasked with stabilizing the chaos are often heavily incentivized to do this. Many a time these folks, in my experience, tend to fail harder and more often than those who try to learn the organizational ropes and tailor their leadership style to fit the organizational culture.*

So don't just jump into your new role. Take the time to talk with as many people as you can. Figure out how to get connected, how to know things, and how to be in the right rooms. Learn the shadow org chart. Solve some problems, but be humble and assume there were good reasons for previous technical decisions—everything has trade-offs. Figure out how to level up the engineers around you. *Understand what's important.* And enjoy a relatively quiet time graph for a while before your calendar fills up with meetings.

Your Choices Matter

We're at the end of the book now! I've got one final thing to say about choosing your path. It's not really about your career, but it's about your job as a senior person in the industry:

You need to take software seriously.

Making software can be fun. There tends to be room for creativity and some whimsy, and most of us don't wear suits to work. But software has a massive influence on everyone's lives. When an application crashes and loses someone's half-written essay, or poor input validation drops someone's health insurance claim, we're wasting people's time and causing them anxiety and stress. The risks of AI and algorithmic bias are well documented. Abuse on social networks, leaks of private information, and deliberately addictive apps destroy people's lives. Our choices can mean that real people suffer.

Software is used for life-critical systems, and that's going to become more common every year. The engineers you level up today may later be responsible for planes, medical treatments, or nuclear power plants. We need to teach new engineers the values of diligence and care that are hallmarks of some other life-critical engineering disciplines. Canadian engineers famously wear a faceted ring (*https://oreil.ly/uEw7I*) intended to remind them of the obligations and ethics of their profession. We need, as an industry, to have the same kind of mindset.

In a time where software engineers are considered *senior* after four to five years, we might forget that there's more to learn. In one where many engineers move on every two to three years, we may be incentivized to build for the current moment (and the current profit and the current promotion) rather than for the long term. Today's college kids and teenagers are going to have enough to deal with: don't send them shoddy systems and technical debt too.

You can take this job seriously and also *really* enjoy it! There's a ton of room for creativity and fun stuff. But bring your good judgment to evaluating the stakes. Know how your software will be used. Be firm about what's negotiable and what's not. You have more influence than you think you do, and the choices you make matter. What the senior people do sets the culture for the industry.

Build good software. Build a good software career. Build a good software industry. Thanks for reading. <3

To Recap

- You are responsible for your career and choices. There are a lot of options about what to optimize for. Know what's important to you. Be deliberate.

- You'll increase your access to opportunities with skills, visibility, relationships, and experiences.

- Everything is learnable if it's worth the time investment.

- Check in with yourself occasionally and make sure your role is still giving you what you need. Look at what's good as well as what's not working.

- There are excellent reasons to spend a long time with one employer. There are excellent reasons to move around too. Either way, you have several options for paths onward.

- Software has a massive influence on the lives and livelihoods of just about everyone on earth. Take the responsibility seriously.

Index

About the Author

Tanya Reilly has over twenty years of experience in software engineering, most recently working on architecture and technical strategy as a senior principal engineer at Squarespace. Previously she was a staff engineer at Google, responsible for some of the largest distributed systems on the planet. Tanya writes about technical leadership and software reliability on her website, No Idea Blog (*https://noidea.dog*). She's an organizer and host of the LeadDev StaffPlus conference and a frequent conference and keynote speaker. Originally from Ireland, she now lives in Brooklyn with her spouse, kid, and espresso machine.

Colophon

The cover paper art is by Susan Thompson. The oak branch in the cover art was created from crepe paper and then photographed. The cover fonts are Gilroy and Guardian Sans. The text fonts are Minion Pro and Scala Pro; the heading and sidebar font is Benton Sans.

Printed in the USA
CPSIA information can be obtained
at www.ICGtesting.com
JSHW010803200324
59436JS00031B/313

9 781098 118730